A Themed Issue of Functional Molecule-Based Magnets

A Themed Issue of Functional Molecule-Based Magnets

Dedicated to Professor Masahiro Yamashita on the Occasion of his 65th Birthday

Special Issue Editor

Keiichi Katoh

MDPI • Basel • Beijing • Wuhan • Barcelona • Belgrade • Manchester • Tokyo • Cluj • Tianjin

Special Issue Editor
Keiichi Katoh
Tohoku University
Japan

Editorial Office
MDPI
St. Alban-Anlage 66
4052 Basel, Switzerland

This is a reprint of articles from the Special Issue published online in the open access journal *Magnetochemistry* (ISSN 2312-7481) (available at: https://www.mdpi.com/journal/magnetochemistry/special_issues/Molecule_based_Magnets).

For citation purposes, cite each article independently as indicated on the article page online and as indicated below:

LastName, A.A.; LastName, B.B.; LastName, C.C. Article Title. *Journal Name* **Year**, *Article Number*, Page Range.

ISBN 978-3-03928-901-1 (Pbk)
ISBN 978-3-03928-902-8 (PDF)

Contents

About the Special Issue Editor

Keiichi Katoh is an assistant professor in the Department of Chemistry of Tohoku University, Japan. He earned his Ph.D. in 2002 from the Graduate University for Advanced Studies, Japan (Institute for Molecular Science), under the supervision of Prof. Katsuya Inoue. He was a post-doctoral fellow with Prof. Takayoshi Nakamura at Hokkaido University, Japan, until 2004 and with Prof. Naoki Sato at Kyoto University, Japan, until 2007. His research focuses on the development of variable spin networks and multi-functional systems based on molecule-based magnetic materials. In April 2020 he will move to Josai University where he is leading research activities in inorganic chemistry and magnetism as an associate professor.

 *magnetochemistry*

Editorial

Special Issue: A Themed Issue of Functional Molecule-Based Magnets: Dedicated to Professor Masahiro Yamashita on the Occasion of His 65th Birthday

Keiichi Katoh

Department of Chemistry, Graduate School of Science, Tohoku University, Aramaki-Aza-Aoba 6-3, Aoba-ku, Sendai 980-8578, Japan; keiichi.katoh.b3@tohoku.ac.jp

Received: 30 March 2020; Accepted: 31 March 2020; Published: 7 April 2020

Research on molecule-based magnetic materials was systematized in the 1980s and expanded rapidly. In Magnetochemistry, a special issue focusing on molecule-based magnetic substances has been published. However, the functionalities of those substances have grown daily, and researchers' quests does not seem to decline. This molecule-based magnetism has developed across many fields, including chemistry, physics, material chemistry, and applied physics, and the use of various functionalities of these molecule-based magnetic substances greatly influences the research on spin-based devices. This special issue is articulated around ten ordinal articles, in honor of Professor Masahiro Yamashita, who contributed greatly to this field, on the occasion of his 65th birthday.

First of all, Barbara Sieklucka will give an introduction to the scholarly achievements of Professor Masahiro Yamashita [1]. He has contributed a huge amount to the study of molecule-based magnets and strongly correlated electron system fields. Shinya Hayami and co-workers have investigated the promise of the pseudo pressure effects of the intercalation spin crossover (SCO) complex by using the transformation of graphene oxide (GO) to reduced graphene oxide (rGO) [2]. Their goal was to provide new insights into the study of pressure effects for molecule-based magnets in two-dimensional (2D) materials, such as graphene, BN, and MoS_2. Takashi Kajiwara, Yasutaka Kitagawa, and co-workers investigated the correlation between the f orbitals of Dy(III) ions and the σ and π orbitals of ligands to bring out the magnetic anisotropy [3]. These results indicated that the coordination geometry and molecular orbitals of Ln complexes should be considered when designing single-molecule magnets (SMMs). Ryuta Ishikawa and co-workers demonstrated the field-induced SMM properties of dinuclear Ln Kramers (Er and Yb) complexes with an electroactive chloralilate-bridging ligand [4]. Their goal was to able to reversibly switch SMM properties via an electrochemical approach. Dawid Pinkowicz and co-workers demonstrated that they could switch the magnetic properties of a Ni(II) coordination polymer under increased pressure [5]. This result indicates that mechanical force can be positively used for changing magnetic properties without physical collapse. Samia Benmansour, Carlos J. Gómez-García, and co-workers investigated the 2D structures of an MnCr honeycomb and observed long-range ferrimagnetic ordering upon changing the solvents [6]. These experimental results can be applied to the reversible switching of T_C (magnetic ordering temperature) by introducing and removing guest molecules. Constantina Papatriantafyllopoulou and co-workers investigated Co_2Ln clusters, which have a triangular metal geometry, and the Co_2Dy derivative exhibited an ac frequency dependence [7]. Mixed $3d/4f$ clusters might provide potential for SMM properties by using a uniaxial magnetic control. Vassilis Psycharis, Mark M. Turnbull, Spyros P. Perlepes, and co-workers demonstrated that the choice of ligand brings about the ability to form interesting structural types in $3d$ metal chemistry [8]. Their work demonstrated the flexibility of ligands and their utility in the synthesis of $3d$ metal complexes with interesting structures and properties. Keiichi Katoh, Masahiro Yamashita, and co-workers investigated the relationship between the coordination geometry and magnetic relaxation phenomena for dinuclear

Dy complexes [9]. These results demonstrate that precise control of the coordination environment enables control of the magnetic relaxation properties. Yasutaka Kitagawa, Masayoshi Nakano, and co-workers examined the possibility of intramolecular magnetic interactions in pyrazole-bridged dinuclear *3d* metal complexes [10]. These results suggest that rational design guidelines for molecule-based magnets based on the quantum chemical calculation would be effective for modifying the interactions and properties of molecule-based magnetic materials.

I have put together a special issue on molecule-based magnets, "A Themed Issue of Functional Molecule-based Magnets: Dedicated to Professor Masahiro Yamashita on the Occasion of his 65th Birthday", and hope it will be of use to everyone. I wish to thank the authors for providing such impressive and interesting papers, and the referees and editorial staff who took the time to write valuable comments.

Author Contributions: K.K. wrote this editorial. The author has read and agreed to the published version of the manuscript.

Funding: This research received no external funding.

Conflicts of Interest: The authors declare no conflict of interest.

References

1. Sieklucka, B. Laudation: In Celebration of Masahiro Yamashita's 65th Birthday. *Magnetochemistry* **2019**, *5*, 25. [CrossRef]
2. Kitayama, H.; Akiyoshi, R.; Nakamura, M.; Hayami, S. Pressure Effects with Incorporated Particle Size Dependency in Graphene Oxide Layers through Observing Spin Crossover Temperature. *Magnetochemistry* **2019**, *5*, 26. [CrossRef]
3. Kobayashi, K.; Harada, Y.; Ikenaga, K.; Kitagawa, Y.; Nakano, M.; Kajiwara, T. Correlation between Slow Magnetic Relaxations and Molecular Structures of Dy(III) Complexes with N5O4 Nona-Coordination. *Magnetochemistry* **2019**, *5*, 27. [CrossRef]
4. Ishikawa, R.; Michiwaki, S.; Noda, T.; Katoh, K.; Yamashita, M.; Kawata, S. Series of Chloranilate-Bridged Dinuclear Lanthanide Complexes: Kramers Systems Showing Field-Induced Slow Magnetic Relaxation. *Magnetochemistry* **2019**, *5*, 30. [CrossRef]
5. Handzlik, G.; Sieklucka, B.; Tomkowiak, H.; Katrusiak, A.; Pinkowicz, D. How to Quench Ferromagnetic Ordering in a CN-Bridged Ni(II)-Nb(IV) Molecular Magnet? A Combined High-Pressure Single-Crystal X-Ray Diffraction and Magnetic Study. *Magnetochemistry* **2019**, *5*, 33. [CrossRef]
6. Martínez-Hernández, C.; Benmansour, S.; Gómez-García, C.J. Chloranilato-Based Layered Ferrimagnets with Solvent-Dependent Ordering Temperatures. *Magnetochemistry* **2019**, *5*, 34. [CrossRef]
7. Efthymiou, C.G.; Ní Fhuaráin, Á.; Mayans, J.; Tasiopoulos, A.; Perlepes, S.P.; Papatriantafyllopoulou, C. A Novel Family of Triangular CoII2LnIII and CoII2YIII Clusters by the Employment of Di-2-Pyridyl Ketone. *Magnetochemistry* **2019**, *5*, 35. [CrossRef]
8. Pilichos, E.; Spanakis, E.; Maniaki, E.-K.; Raptopoulou, C.P.; Psycharis, V.; Turnbull, M.M.; Perlepes, S.P. Diversity of Coordination Modes in a Flexible Ditopic Ligand Containing 2-Pyridyl, Carbonyl and Hydrazone Functionalities: Mononuclear and Dinuclear Cobalt(III) Complexes, and Tetranuclear Copper(II) and Nickel(II) Clusters. *Magnetochemistry* **2019**, *5*, 39. [CrossRef]
9. Sato, T.; Matsuzawa, S.; Katoh, K.; Breedlove, B.K.; Yamashita, M. Relationship between the Coordination Geometry and Spin Dynamics of Dysprosium(III) Heteroleptic Triple-Decker Complexes. *Magnetochemistry* **2019**, *5*, 65. [CrossRef]
10. Fujii, T.; Kitagawa, Y.; Ikenaga, K.; Tada, H.; Era, I.; Nakano, M. Theoretical Study on Magnetic Interaction in Pyrazole-Bridged Dinuclear Metal Complex: Possibility of Intramolecular Ferromagnetic Interaction by Orbital Counter-Complementarity. *Magnetochemistry* **2020**, *6*, 10. [CrossRef]

 *magnetochemistry*

Editorial

Laudation: In Celebration of Masahiro Yamashita's 65th Birthday

Barbara Sieklucka

Faculty of Chemistry, Jagiellonian University, Gronostajowa 2, 30-387 Krakow, Poland; barbara.sieklucka@uj.edu.pl

Received: 3 April 2019; Accepted: 3 April 2019; Published: 10 April 2019

Professor Masahiro Yamashita at the Tohoku University, Japan, celebrates his 65th birthday in 2019. His co-workers, colleagues, and friends congratulate him on this happy occasion. For the celebration, the Special Issue of "A Themed Issue of Functional Molecule-Based Magnets: Dedicated to Professor Masahiro Yamashita on the Occasion of his 65th Birthday" encompasses articles submitted by authors from all around the globe. The range of international contributions reflects Masahiro's diverse scientific interests as well as his close relationship with the molecular magnetism community.

Professor Masahiro Yamashita received his Doctor of Science in 1982 from the Kyushu University. He then joined the Institute for Molecular Science (IMS). In 1985, he was appointed to the position of Assistant Professor at Kyushu University. In 1989, he was appointed to the position of Associate Professor at Nagoya University. He was a Full Professor at Tokyo Metropolitan University from 2000 to 2004. He is now a Full Professor at Tohoku University and Principal Investigator of the Soft Materials Group in Advanced Institute for Materials Research (AIMR) at Tohoku University.

He is also Visiting Professor at Nanjing University, Zhenjiang University, Xi'an Jaotong University, Sun Yat Sen University, Peking University, Guilin Normal University, Nankai University (China), and Cagliari University (Italy).

Professor Masahiro Yamashita is one of the most influential scientists in the fields of multifunctional nanoscience of advanced metal complexes. In his research he has focused on quantum molecular spintronics, single molecule and single chain quantum magnets, strongly electron correlated nanowire metal complexes, and molecule-based magnets encapsulated in carbon nanotubes.

He has been honored with the following awards: the Inoue Scientific Award (2002), the Chemical Society of Japan Award for Creative Work (2005), the Award of Japan Society of Coordination Chemistry (2014), and the Mukai Award (2019). He is now an Associate Member of the Science Council of Japan. He is also Associate Editor of Dalton Transactions as well as a Fellow of the Royal Society of Chemistry (FRSC). He is known for his strong dedication to research, teaching, and mentoring of students. In the course of his academic career he has guided 47 Ph.D. students and numerous masters students, through to graduation.

He has successfully organized the most prestigious conferences in the field of molecular magnetism which include: the International Conference, "Single-Molecule Quantum Magnets and Single-Chain Quantum Magnets—New Generation of Quantum Nanomagnets", Okazaki, Japan, 2006; the 62nd Fujihara Seminar, "Frontier and Perspectives in Molecule-Based Quantum Magnets", Sendai, Japan, 2012; as well as the 15th International Conference on Molecule-Based Magnets (ICMM2016) and the 43rd International Conference on Coordination Chemistry (ICCC2018) in Sendai, Japan.

Professor Masahiro Yamashita's scientific activity is evidenced by ca. 450 research articles, 90 reviews, 20 books and book chapters, and countless contributions at conference lectures.

On this special occasion, I would like to join Masahiro's students, co-workers, collaborators, and friends in congratulating him on this 65th birthday. We all wish for him to continue and to enjoy his scientific endeavors in good health.

Funding: This research received no external funding.

Conflicts of Interest: The author declares no conflict of interest.

 *magnetochemistry*

Article

Pressure Effects with Incorporated Particle Size Dependency in Graphene Oxide Layers through Observing Spin Crossover Temperature

Hikaru Kitayama [1], Ryohei Akiyoshi [1], Masaaki Nakamura [1] and Shinya Hayami [1,2,*]

[1] Department of Chemistry, Graduate School of Science and Technology, Kumamoto University, 2-39-1 Kurokami, Chuo-ku, Kumamoto 860-8555, Japan; 179d8024@st.kumamoto-u.ac.jp (H.K.); 187d9041@st.kumamoto-u.ac.jp (R.A.); m_nakamura@kumamoto-u.ac.jp (M.N.)
[2] Institute of Pulsed Power Science (IPPS), Kumamoto University, 2-39-1 Kurokami, Chuo-ku, Kumamoto 860-8555, Japan
* Correspondence: hayami@kumamoto-u.ac.jp; Tel.: +81-096-342-3469

Received: 23 March 2019; Accepted: 8 April 2019; Published: 11 April 2019

Abstract: This research highlights the pressure effects with the particle size dependency incorporated in two-dimensional graphene oxide (GO)/reduced graphene oxide (rGO). GO and rGO composites employing nanorods (NRs) of type [Fe(Htrz)$_2$(trz)](BF$_4$) have been prepared, and their pressure effects in the interlayer spaces through observing the changes of the spin crossover (SCO) temperature ($T_{1/2}$) have been discussed. The composites show the decrease of interlayer spaces from 8.7 Å to 3.5 Å that is associated with GO to rGO transformation. The shorter interlayer spaces were induced by pressure effects, resulting in the increment of $T_{1/2}$ from 357 K to 364 K. The pressure effects in the interlayers spaces estimated from the $T_{1/2}$ value correspond to 24 MPa in pristine [Fe(Htrz)$_2$(trz)](BF$_4$) NRs under hydrostatic pressure. The pressure observed in the composites incorporating NRs (30 × 200 nm) is smaller than that observed in the composite incorporating nanoparticles (NPs) (30 nm). These results clearly demonstrated that the incorporated particle size and shape influenced the pressure effects between the GO/rGO layer.

Keywords: pressure effect; spin crossover; graphene oxide; iron complex

1. Introduction

Van der Waals interactions in the pores of micro-porous materials are known to generate a pseudo-pressure effect, leading to the expression of characteristic phases and unique properties in the pores under mild conditions. [1–6]. For example, potassium iodide (KI) nanocrystals inside the nanotube spaces of single-walled carbon nano-horns display a structural phase transition by pseudo-pressure corresponding to *ca*. 1.9 GPa [7]. In microporous of metal-organic frameworks [{[Cu$_2$(pzdc)$_2$(pyz)]·2H$_2$O}$_n$] (pzdc—pyrazine-2,3-dicarboxylate), O$_2$ molecules show similar behavior to the solid phase above the freezing point of O$_2$ [8].

Recently, interlayers of two-dimensional (2D) materials, such as graphene, boron nitride (BN), and MoS$_2$, were found to play an important role for the confinement of molecules and pseudo-pressure effects [9–12]. For instance, pressure corresponding to 1.2 ± 0.3 GPa was observed by trapping pressure-sensitive molecules of triphenyl amine (TPA) and boric acid (BA) into an interlayer of graphene [9]. In typical 2D layered materials, the correlations between pressure (P) and the interlayer distance (d) were estimated using the equation of $P \approx E_w/d$, where E_w is the adhesion energy [11,12]. As such this is an indication that the pressure effects that occur in the interlayer are significantly affected by the interlayer distance. Thus, 2D materials that possess a tunable interlayer have the possibility of tuning pressure effects, leading to the generation of unique phases and physical properties.

Graphene oxide (GO), an oxidized graphene, is a 2D material that has oxygen functional groups, such as hydroxyl, carboxyl, and epoxy groups [13–16]. These oxygen functional groups on the GO surface were removed by thermal reduction treatment, resulting in reduced graphene oxide (rGO) [17–19]. Importantly, their interlayer distances decrease from 7.9 Å in GO to 3.4 Å in rGO as a result of the removal of the oxygen functional groups [20–22]. Therefore, a pseudo pressure effect can be generated via GO/rGO transformation.

Recently, we reported tunable pressure effects on GO/rGO layers by changing the thermal treatment temperature [23]. In this context, nanoparticles (NPs) of a spin crossover (SCO) complex of type $[Fe(Htrz)_2(trz)](BF_4)$ (trz: 1,2,4-triazole) were used for monitoring the pressure effect changes on the GO/rGO layers. SCO complexes, $[Fe(Htrz)_2(trz)](BF_4)$, are also well known to exhibit SCO phenomena between low spin (LS) and high spin (HS) states, reversibly with thermal hysteresis [24–27]. In addition, the SCO temperature ($T_{1/2}$) of $[Fe(Htrz)_2(trz)](BF_4)$ is sensitive to the hydrostatic pressure that behaves to restrict the structural transition synchronized with the SCO behavior. The correlation between P and $T_{1/2}$ has also been reported in $[Fe(Htrz)_2(trz)](BF_4)$ NPs [28]. In a prior study, we reported that the composite incorporating $[Fe(Htrz)_2(trz)](BF_4)$ bulk particles (BPs) of 100 nm did not show any pressure effect, but did exhibit a pressure effect for the composite incorporating NPs of 30 nm, since the GO nanosheet can cover the NPs completely.

In the present study, we aimed to further investigate the pressure effects between the GO/rGO layers. For this purpose, we prepared GO (**1**)/rGO (**2**) composites incorporating cylinder shape nanorods (NRs) $[Fe(Htrz)_2(trz)](BF_4)$ with a size of 30 × 200 nm (intermediate size of previously reported particle) as a way to detect the pressure effects (Figure 1). Then, pressure effects in the GO/rGO layers were discussed by monitoring $T_{1/2}$, and a comparison was made with GO/rGO composites incorporating spherical NPs (30 × 30 nm) and bulk particles (100 × 100 nm) respectively.

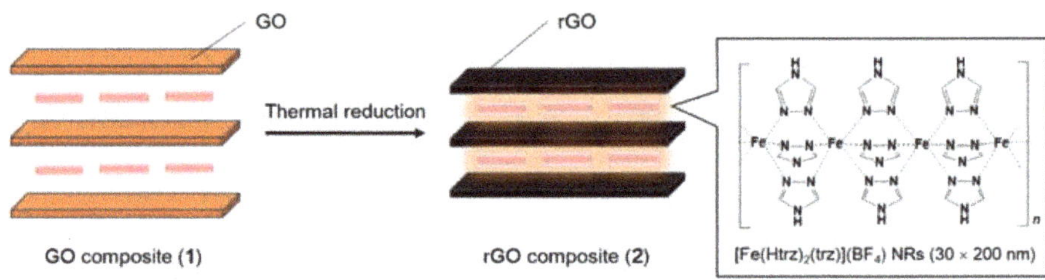

Figure 1. Schematic illustration of pressure effects in graphene oxide (GO)/ reduced graphene oxide (rGO) layers incorporating $[Fe(Htrz)_2(trz)](BF_4)$ nanorods (NRs).

2. Results and Discussion

The $[Fe(Htrz)_2(trz)](BF_4)$ NRs were synthesized by the reaction between $FeCl_2·4H_2O$, $NaBF_4$, and 1-H-1,2,4-triazole, using the ligand-melt method [29]. Composite **1** was prepared by mixing GO and $[Fe(Htrz)_2(trz)](BF_4)$ NRs in a mass ratio of 1:2 in ethanol, which was then filtrated. Composite **2** was obtained by subsequent heating at 473 K for 12 h. The GO/rGO transformation in these composites was confirmed by investigating the current–voltage (I–V) properties. The I–V curve for composite **1** shows mainly an insulator property in accordance with the behavior of GO. The electron conductivity of composite **1** was $7.67 × 10^{-11}$ A, applied at 1 V. On the other hand, composite **2** showed $7.28 × 10^{-6}$ A applied at 1 V, in accordance with the oxygen functional groups being removed to yield rGO. This transformation is also corroborated by the powder X-ray diffraction (PXRD) patterns, as presented in Figure 3.

The scanning electron microscopy (SEM) images of the $[Fe(Htrz)_2(trz)](BF_4)$ NRs, composite **1**, and composite **2** are presented in Figure 2 and Figure S2. The SEM image demonstrated that the size of the NR complex was 29.6 nm in width and 203.4 nm in length. For composites **1** and **2**, the NRs incorporated between the GO/rGO layers were observed obviously. Furthermore, the presence of [Fe(Htrz)$_2$(trz)](BF$_4$) was clearly confirmed by the energy dispersive X-ray (EDX) spectroscopy (Figure 2c,d). The Fourier transform infrared spectra (FT-IR) results also supported the presence of [Fe(Htrz)$_2$(trz)](BF$_4$) NRs composited within the GO/rGO interlayers (Figure S3).

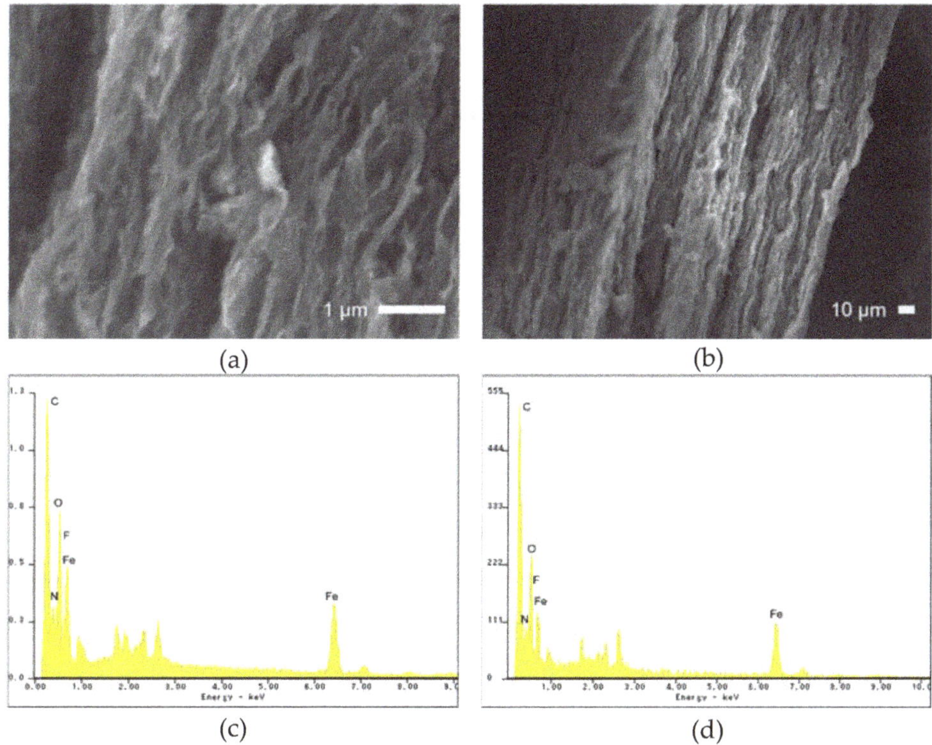

Figure 2. Scanning electron microscopy (SEM) images of (**a**) composite **1** and (**b**) composite **2**. SEM-energy dispersive X-ray spectroscopy (SEM-EDX) results for (**c**) composite **1** and (**d**) composite **2**.

As such, the changes of the interlayer distance that is associated with the transformation of GO to rGO were investigated by powder X-ray diffraction (PXRD) measurements (Figure 3). Results shows that pristine GO has a distinct peak at $2\theta = 10.15°$, with an interlayer distance of 8.70 Å. As for composite **1**, the GO peak was observed at $2\theta = 10.17°$ and an interlayer distance of 8.68 Å, where the remaining peaks are ascribed to the presence of [Fe(Htrz)$_2$(trz)](BF$_4$) NRs. In the case of composite **2** (which was treated at 473 K for 12 h), the interlayer distance decreased to 3.5 Å ($2\theta = 25°$) as a result of the removal of the oxygen functional groups on the GO layers. From these results, it can be anticipated that pressure effects occurred between the interlayers.

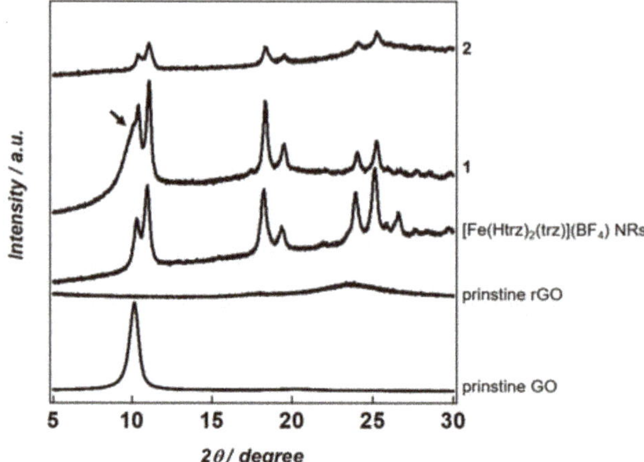

Figure 3. Powder X-ray diffraction (PXRD) patterns for pristine GO, pristine rGO, [Fe(Htrz)$_2$(trz)](BF$_4$) NRs, composite **1**, and composite **2**.

In order to investigate the influence of the pressure effects on the SCO behavior, caused by the shorter interlayer distance associated with the structural transformation between GO and rGO, the temperature-dependent magnetic susceptibility for the [Fe(Htrz)$_2$(trz)](BF$_4$) NRs, composite **1**, and composite **2** were measured in the temperature range of 300 to 400 K. The magnetic susceptibility for the [Fe(Htrz)$_2$(trz)](BF$_4$) NRs in the form of the $\chi_m T$ vs. T plot can be seen in Figure S4, where χ_m is the molar magnetic susceptibility and T is the temperature. From these results, [Fe(Htrz)$_2$(trz)](BF$_4$) NRs show SCO behavior at $T_{1/2}$ = 356 K, with a thermal hysteresis of 29 K. The $\chi_g T$ vs. T plots for composite **1** and composite **2** are shown in Figure 4, where χ_g is the magnetic susceptibility per gram. Both composites **1** and **2** exhibited SCO behavior at $T_{1/2}$ = 357 K and 364 K respectively. The $T_{1/2}$ value of composite **2** is 7 K higher than that observed in composite **1**. Accordingly, these results are in agreement with pressure effects behavior when decreasing the interlayer distance.

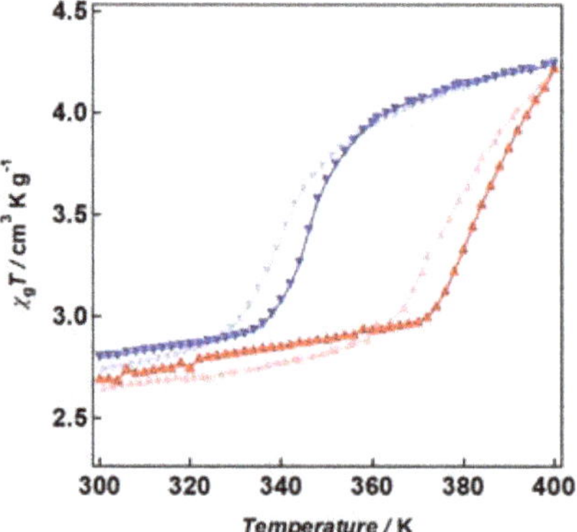

Figure 4. $\chi_g T$ vs. T plots for composite **1** (heating: ( ); cooling: (▼)) and composite **2** (heating: (▲); cooling: (▼)).

The pseudo-pressure effects were estimated from the $T_{1/2}$ value using the Clausius–Clapeyron equation (Equation (1)) reported by Colacio and co-workers, where p is hydrostatic pressures [28], as follows:

$$T_{1/2}(p) = T_{1/2} + 290(66)p \tag{1}$$

The values of SCO temperature and pseudo-pressure for GO/rGO composite when [Fe(Htrz)$_2$(trz)](BF$_4$) of different size and shape are incorporated are summarized in Table 1. As a result of the calculation, the pseudo-pressure originated from the transformation of composite **1** to composite **2** is equal to 24 MPa. We have reported previously that GO/rGO composites incorporating [Fe(Htrz)$_2$(trz)](BF$_4$) NPs with a size of 30 nm show an increase of the $T_{1/2}$ value from $T_{1/2}$ = 351 K in the GO, to $T_{1/2}$ = 362 K in rGO due to pressure effects corresponding to 38 MPa [23]. The pseudo-pressure effect observed in the composite with NRs (30 × 200 nm) was smaller than that observed in the composite with NPs (30 nm). Considering that no pressure effects were observed for the composite incorporating BPs of 100 nm size, it can be concluded that the accommodated particle size and shape crucially affected the pseudo-pressure effects within the GO/rGO layers. For the case of small particle size, the GO layers stack regularly. GO layers form the ordered stacking structures when incorporating NRs, however, the surface area of the NRs influencing the pressure effects is larger than the NPs with a size of 30 nm (Figure 5). It is then proposed that a large surface of NRs leads to small pressure effects.

Table 1. Summary of spin crossover (SCO) temperatures ($T_{1/2}$) and pseudo-pressure.

	$T_{1/2}$ (K)	Pseudo-Pressure
1 (GO/NRs)	357	24 MPa
2 (rGO/NRs)	364	
GO/NPs	351	38 MPa
rGO/NPs	362	
GO/BPs	357	No pressure
rGO/BPs	352	

The particle size is 30 × 200 nm in nanorods (NRs), 30 nm in nanoparticles (NPs), and 100 nm in bulk particles (BPs).

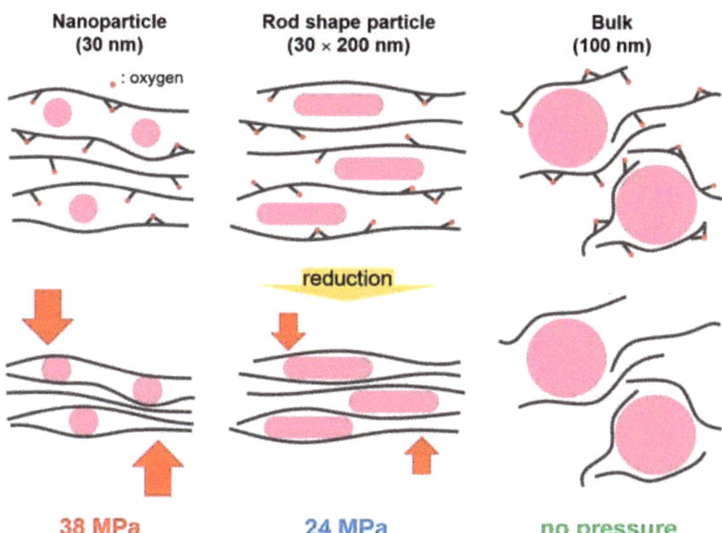

Figure 5. Schematic illustration of the pressure effects with incorporated particle size and shape dependency in the GO/rGO layers.

3. Materials and Methods

3.1. Synthesis

All the materials and reagents were obtained from Wako Pure Chemical Industries (Osaka-shi, Osaka, Japan) and Tokyo Chemical Industry (TCI) Co., Ltd (Chuo-ku, Tokyo, Japan) and used without further purification.

3.1.1. [Fe(Htrz)$_2$(trz)](BF$_4$) NRs (30 × 200 nm)

The [Fe(Htrz)$_2$(trz)](BF$_4$) NRs were prepared according to the previously reported procedure [29]. The mixture of FeCl$_2$·4H$_2$O (200 mg, 1 mmol), NaBF$_4$ (110 mg, 1 mmol), and 1-H-1,2,4-triazole (5 g, 72.4 mmol) was heated at 150 °C for 5 min. After heating, the resulting melt was cooled to room temperature. The obtained crude product was dispersed in ethanol. The dispersion was centrifuged at 4800 r/min, washed with ethanol, and then collected using a membrane filter (1 μm) so as to give the product as a violet powder.

3.1.2. Graphene Oxide (GO)

The graphene oxide was prepared by Hummer's method with a minor modification [21]. The mixture of graphite (2 g), grinded NaNO$_3$ (2 g), and H$_2$SO$_4$ (92 mL) was stirred for 30 min at 0 °C. Subsequently, KMnO$_4$ powder (10 g) was added carefully, and the resulting mixture was stirred at 35 °C for 60 min. Then, deionized water (92 mL) was dropped into the mixture slowly, and the mixture was heated at 95 °C for 20 min. Subsequently, deionized water (200 mL) was poured into the reaction mixture. Then, a 30% H$_2$O$_2$ solution (30 mL) was dropped very carefully so as to convert the manganese dioxide and unreacted permanganate into soluble sulfates in an ice bath. The mixture was centrifuged at 3000 r/min to remove the supernatant liquid. The precipitate was washed with a 5% HCl solution three times, and then with distilled water five times. The resulting solid was washed with ionized water three times, and exfoliated by ultrasonication for 2 h. The solution was centrifuged at 8000 r/min for 30 min, then the supernatant dispersion was centrifuged at 15000 r/min for 30 min to give the graphene oxide (GO) dispersion.

3.1.3. GO–[Fe(Htrz)$_2$(trz)](BF$_4$) NRs Composite (**1**)

The mixture of GO in ethanol (30 mg/50 mL) and [Fe(Htrz)$_2$(trz)](BF$_4$) NRs in ethanol (60 mg/50 mL) was stirred at 25 °C for 6 h. After stirring, the brown product was centrifuged at 4000 r/min for 30 min, and the crude product was collected using a membrane filter (1 μm), washed with ethanol to give the product.

3.1.4. rGO–[Fe(Htrz)$_2$(trz)](BF$_4$) NRs Composite (**2**)

Composite **1** was reduced to composite **2** by thermal treatments in a vacuum at 473 K for 12 h.

3.2. Measurement

All the measurements for composites **1** and **2** were performed three times using a film sample. Current-voltage (IV) properties were measured using an electrochemical analyzer, BAS, Model ALS/DY2323 BI-POTENTIOSTAT (Sumida-ku, Tokyo, Japan). The scanning electron microscopy (SEM) images and SEM-energy dispersive X-ray spectroscopy (SEM-EDX) data were collected on a JEOL, JSM-7600 F instrument (Akishima-shi, Tokyo, Japan). Fourier transform infrared spectra (FT-IR) were collected on SHIMADZU, IRAAffinity-1S (Kyoto-shi, Kyoto, Japan). The powder X-ray diffraction (PXRD) patterns were recorded on a Rigaku, MiniFlex II X-ray diffractometer (Akishima-shi, Tokyo, Japan). The temperature dependence of magnetic susceptibilities was measured on a Superconducting Quantum Interference Device (SQUID) magnetometer, Quantum Design Japan,

MPMSXL-5 (Toshima-ku, Tokyo, Japan). The samples were placed inside the SQUID chamber and measured between 300 and 400 K at field strengths of 0.5 T.

4. Conclusions

In summary, we have prepared a composite consisting of GO/rGO and a SCO complex of $[Fe(Htrz)_2(trz)](BF_4)$ NRs (30 × 200 nm). It was found that the shorter interlayer caused by the transformation of GO to rGO leads to pseudo-pressure effects. GO composite exhibited SCO behavior at $T_{1/2} = 357$ K, whereas for rGO composite $T_{1/2}$ was at 364 K. The observed pressure for $[Fe(Htrz)_2(trz)](BF_4)$ NRs estimated from the $T_{1/2}$ value corresponded to 24 MPa and being lower than that observed for the case of NPs (30 nm). Clearly, these findings provide new insight towards the research regarding the pressure effects in 2D materials including graphene, BN, MoS_2.

Supplementary Materials: The following are available online at http://www.mdpi.com/2312-7481/5/2/26/s1. Figure S1: I–V curves for **1** and **2**. Figure S2: SEM image of $[Fe(Htrz)_2(trz)](BF_4)$ NRs. Figure S3: FT-IR spectra for $[Fe(Htrz)_2(trz)](BF_4)$ NRs, composite **1**, and composite **2**. Figure S4: $\chi_m T$ vs. T plot for $[Fe(Htrz)_2(trz)](BF_4)$ NRs. Figure S5: $\chi_g T$ vs T plots for composite **1** and composite **2**.

Author Contributions: The manuscript was written through the contributions of all of the authors. All of the authors have given approval to the final version of the manuscript. The work was directed by S.H.

Funding: This work was supported by KAKENHI Grant-in-Aid for Scientific Research (A) JP17H01200.

Conflicts of Interest: The author declares no conflict of interest.

References

1. Granick, S. Motions and Relaxations of Confined Liquids. *Science* **1991**, *253*, 1374–1379. [CrossRef] [PubMed]
2. Hashimoto, S.; Fujimori, T.; Tanaka, H.; Urita, K.; Ohba, T.; Kanoh, H.; Itoh, T.; Asai, M.; Sakamoto, H.; Niimura, S.; et al. Anomaly of CH_4 Molecular Assembly Confined in Single-Wall Carbon Nanohorn Spaces. *J. Am. Chem. Soc.* **2011**, *133*, 2022–2024. [CrossRef] [PubMed]
3. Iiyama, T.; Nishikawa, K.; Otowa, T.; Kaneko, K. An Ordered Water Molecular Assembly Structure in a Slit-Shaped Carbon Nanospace. *J. Phys. Chem.* **1995**, *99*, 10075–10076. [CrossRef]
4. Casco, M.E.; Silvestre-Albero, J.; Ramírez-Cuesta, A.J.; Rey, F.; Jordá, J.L.; Bansode, A.; Urakawa, A.; Peral, I.; Martínez-Escandell, M.; Kaneko, K.; et al. Methane hydrate formation in confined nanospace can surpass nature. *Nat. Commun.* **2015**, *6*, 6432–6439. [CrossRef] [PubMed]
5. Sun, L.; Banhart, F.; Krasheninnikov, A.V.; Rodríguez-Manzo, J.A.; Terrones, M.; Ajayan, P.M. Carbon Nanotubes as High-Pressure Cylinders and Nanoextruders. *Science* **2006**, *312*, 1199–1202. [CrossRef] [PubMed]
6. Uemura, T.; Yanai, N.; Watanabe, S.; Tanaka, H.; Numaguchi, R.; Miyahara, M.T.; Ohta, Y.; Nagaoka, M.; Kitagawa, S. Unveiling thermal transitions of polymers in subnanometre pores. *Nat. Commun.* **2010**, *1*, 83–90. [CrossRef]
7. Urita, K.; Shiga, Y.; Fujimori, T.; Iiyama, T.; Hattori, Y.; Kanoh, H.; Ohba, T.; Tanaka, H.; Yudasaka, M.; Iijima, S.; et al. Confinement in Carbon Nanospace-Induced Production of KI Nanocrystals of High-Pressure Phase. *J. Am. Chem. Soc.* **2011**, *133*, 10344–10347. [CrossRef]
8. Kitamura, R.; Kitagawa, S.; Kubota, Y.; Kobayashi, T.C.; Kindo, K.; Mita, Y.; Matsuo, A.; Kobayashi, M.; Chang, H.-C.; Ozawa, C.T.; et al. Formation of a One-Dimensional Array of Oxygen in a Microporous Metal-Organic Solid. *Science* **2002**, *298*, 2358–2361. [CrossRef] [PubMed]
9. Vasu, K.S.; Prestat, E.; Abraham, J.; Dix, J.; Kashtiban, R.J.; Beheshtian, J.; Sloan, J.; Carbone, P.; Neek-Amal, M.; Haigh, S.J.; et al. Van der Waals pressure and its effect on trapped interlayer molecules. *Nat. Commun.* **2016**, *7*. [CrossRef]
10. Chialvo, A.A.; Vlcek, L.; Cummings, P.T. Surface Strain Effects on the Water-Graphene Interfacial and Confinement Behavior. *J. Phys. Chem. C* **2014**, *118*, 19701–19711. [CrossRef]
11. Khestanova, E.; Guinea, F.; Fumagalli, L.; Geim, A.K.; Grigorieva, I.V. Universal shape and pressure inside bubbles appearing in van der Waals heterostuctures. *Nat. Commun.* **2016**, *7*. [CrossRef] [PubMed]
12. Algara-Siller, G.; Lehtinen, O.; Wang, F.C.; Nair, R.R.; Kaiser, U.; Wu, H.A.; Geim, A.K.; Grigorieva, I.V. Square ice in graphene nanocapollaries. *Nature* **2015**, *519*, 443–445. [CrossRef] [PubMed]

13. Karim, M.R.; Takehira, H.; Matsui, T.; Murashima, Y.; Ohtani, R.; Nakamura, M.; Hayami, S. Graphene and Graphene Oxide as Super Materials. *Curr. Inorg. Chem.* **2014**, *4*, 191–219. [CrossRef]

14. Karim, M.R.; Hatakeyama, K.; Matsui, T.; Takehira, H.; Taniguchi, T.; Koinuma, M.; Matsumoto, Y.; Akutagawa, T.; Nakamura, T.; Noro, S.; et al. Graphene Oxide Nanosheet with High Proton Conductivity. *J. Am. Chem. Soc.* **2013**, *135*, 8079–8100. [CrossRef] [PubMed]

15. Dreyer, D.R.; Park, S.; Bielawski, C.W.; Ruoff, R.S. The chemistry of graphene oxide. *Chem. Soc. Rev.* **2010**, *39*, 228–240. [CrossRef] [PubMed]

16. Morimoto, N.; Suzuki, H.; Takeuchi, Y.; Kawaguchi, S.; Kunisu, M.; Bielawski, C.W.; Nishina, Y. Real-Time, in Situ Monitoring of the Oxidation of Graphite: Lessons Learned. *Chem. Mater.* **2017**, *29*, 2150–2156. [CrossRef]

17. Pei, S.; Cheng, H.-M. The reduction of graphene oxide. *Carbon* **2012**, *50*, 3210–3228. [CrossRef]

18. Eda, G.; Chhowalla, M. Chemically Derived Graphene Oxide: Towards Large-Area Thin-Film Electronics and Optoelectronics. *Adv. Mater.* **2010**, *22*, 2392–2415. [CrossRef]

19. Hatakeyama, K.; Tateishi, H.; Taniguchi, T.; Koinuma, M.; Kida, T.; Hayami, S.; Yokoi, H.; Matsumoto, Y. Tunable Graphene Oxide Proton/Electron Mixed Conductor that Functions at Room Temperature. *Chem. Mater.* **2014**, *26*, 5598–5604. [CrossRef]

20. Pan, Q.; Chung, C.-C.; He, N.; Jones, J.L.; Gao, W. Accelerated Thermal Decomposition of Graphene Oxide Films in Air via in Situ X-ray Diffraction Analysis. *J. Phys. Chem. C* **2016**, *120*, 14984–14990. [CrossRef]

21. Shin, H.-J.; Kim, K.K.; Benayad, A.; Yoon, S.-M.; Park, H.K.; Jung, I.-S.; Jin, M.H.; Jeong, H.-K.; Kim, J.M.; Choi, J.-Y.; et al. Efficient Reduction of Graphite Oxide by Sodium Borohydride and Its Effect on Electrical Conductance. *Adv. Funct. Mater.* **2009**, *19*, 1987–1992. [CrossRef]

22. Tien, H.-W.; Huang, Y.-L.; Yang, S.-Y.; Wang, J.-Y.; Ma, C.-C.M. The production of graphene nanosheets decorated with silver nanoparticles for use in transparent, conductive films. *Carbon* **2011**, *49*, 1550–1560. [CrossRef]

23. Sekimoto, Y.; Ohtani, R.; Nakamura, M.; Koinuma, M.; Lindoy, L.F.; Hayami, S. Tunable pressure effects in graphene oxide layers. *Sci. Rep.* **2017**, *7*, 1–7. [CrossRef]

24. Kröber, J.; Audière, J.-P.; Claude, R.; Codjovi, E.; Kahn, O. Spin Transitions and Thermal Hystereses in the Molecular-Based Materials[Fe(Htrz)$_2$(trz)](BF$_4$)$_2$ and [Fe(Htrz)$_3$](BF$_4$)$_2$·H$_2$O (Htrz = 1,2,4-4H-triazole; trz = 1,2,4-triazolato). *Chem. Mater.* **1994**, *6*, 1404–1412. [CrossRef]

25. Coronado, E.; Galán-Mascarós, J.R.; Monrabal-Capilla, M.; García-Martínez, J.; Pardo-Ibáñez, P. Bistable Spin-Crossover Nanoparticles Showing Magnetic Thermal Hysteresis near Room Temperature. *Adv. Mater.* **2007**, *19*, 1359–1361. [CrossRef]

26. Kahn, O.; Codjovi, E. Iron(II)-1,2,4-triazole spin transition molecular materials. *Philos. Trans. R. Soc. Lond. A* **1996**, *354*, 359–379. [CrossRef]

27. Holovchenko, A.; Dugay, J.; Giménez-Maequés, M.; Torres-Cavanillas, R.; Coronado, E.; van der Zant, H.S.J. Near Room-Temperature Memory Devices Based on Hybrid Spin-Crossover@SiO2 Nanoparticles Coupled to Single-Layer Graphene Nanoelectrodes. *Adv. Mater.* **2016**, *28*, 7228–7233. [CrossRef] [PubMed]

28. Herrera, J.M.; Titos-Padilla, S.; Pope, S.J.A.; Berlanga, I.; Zamora, F.; Delgado, J.J.; Kamenev, K.V.; Wang, X.; Prescimone, A.; Brechin, E.K.; et al. Studies on bifunctional Fe(II)-triazole spin crossover nanoparticles: time-dependent luminescence, surface grafting and the effect of a silica shell and hydrostatic pressure on the magnetic properties. *J. Mater. Chem. C* **2015**, *3*, 7819–7829. [CrossRef]

29. Rotaru, A.; Cural'skiy, I.A.; Molnár, G.; Salmon, L.; Demont, P.; Bousseksou, A. Spin state dependence of electrical conductivity of spin crossover materials. *Chem. Commun.* **2012**, *48*, 4163–4165. [CrossRef]

 *magnetochemistry*

Article

Correlation between Slow Magnetic Relaxations and Molecular Structures of Dy(III) Complexes with N_5O_4 Nona-Coordination

Kaede Kobayashi [1], Yukina Harada [1], Kazuki Ikenaga [2], Yasutaka Kitagawa [2,*], Masayoshi Nakano [2,3] and Takashi Kajiwara [1,*]

[1] Department of Chemistry, Biology, and Environmental Science, Faculty of Science, Nara Women's University, Kita-uoya Nishi-machi, Nara 630-8506, Japan; sak_kobayashi@cc.nara-wu.ac.jp (K.K.); yh3402@gmail.com (Y.H.)

[2] Department of Materials Engineering Science, Graduate School of Engineering Science, Osaka University, Toyonaka, Osaka 560-8531, Japan; kazuki.ikenaga@cheng.es.osaka-u.ac.jp (K.I.); mnaka@cheng.es.osaka-u.ac.jp (M.N.)

[3] Institute for Molecular Science, Myodaiji, Okazaki 444-8585, Japan

* Correspondence: kitagawa@cheng.es.osaka-u.ac.jp (Y.K.); kajiwara@cc.nara-wu.ac.jp (T.K.); Tel.: +81-6-6850-6267 (Y.K.); +81-742-20-3402 (T.K.)

Received: 30 March 2019; Accepted: 13 April 2019; Published: 18 April 2019

Abstract: A series of Dy(III) mononuclear complexes $[DyA_2L]^+$ (L denotes Schiff base N_5 ligand that occupies equatorial positions and A^- denotes bidentate anionic O-donor ligands such as NO_3^- (**1**), AcO^- (**2**), and $acac^-$ (**3**)) were synthesized to investigate the correlation between the slow magnetic relaxation phenomena and the coordination structures around Dy(III). The Dy(III) ion in each complex is in a nona-coordination with the anionic O-donor ligand occupying up- and down-side positions of the N_5 equatorial plane. **2** and **3** show slow magnetic relaxation phenomena under a zero bias-field condition, and all complexes showed slow magnetic relaxation under the applied 1000-Oe bias-field conditions. Arrhenius analyses revealed that the $\Delta E/k_B$, the barrier height for magnetization flipping, increases in this order, with the values of 24.1(6), 85(3), and 140(15) K. The effects of the exchanging axial ligands on the magnetic anisotropy were discussed together with the DFT calculations.

Keywords: lanthanide complex; slow magnetic relaxation; single-molecule magnet; crystal structure; AC susceptibility; DFT calculation

---

1. Introduction

Single-molecule magnets (SMMs) are fascinating molecule-based nanomaterials, which are characterized by slow relaxation of magnetization at low temperatures [1–10]. Magnetic anisotropy plays an essential role in preventing the magnetization flipping; as the orbital angular momentum of 4f electrons is unquenched in the complex formations, each lanthanide(III) (Ln(III)) ion possesses a large magnetic moment correlated with the total angular momentum J, which is defined by the length of the vector summation of the spin angular momentum S and the orbital angular momentum L. The Dy(III) ion is the most fascinating lanthanide ion due to a large total angular momentum of $J = 15/2$, accompanied by the Kramers characteristic and an oblate type electronic distribution [11–19]. The magnetic anisotropy of lanthanide ions is strongly correlated with the electronic repulsion with the crystal field of an appropriate anisotropy; this means an axially stressed crystal field is advantageous for realizing an easy axis anisotropy of the oblate type lanthanide ion. The J ground state splits into $2J + 1$ numbered substates with different components along the z axis, i.e., J_z. In an appropriate anisotropic crystal field, the pair of substates with the highest Ising character was relatively stabilized compared with those of less Ising type pairs to give an easy axial magnetic anisotropy. To achieve such an anisotropic crystal field, the combination of neutral and anionic ligands, with the former located at equatorial positions and the latter located along the *z* axis, is simple but powerful [18,20–23]. This strategy is effective both for light and heavy lanthanide ions with oblate type electronic distributions, such as Ce(III), Nd(III) [24,25], Tb(III), and Dy(III). To this effect, we have previously reported several mononuclear Ln(III) complexes (Ln = Ce, Pr, and Nd) incorporated with neutral and anionic ligand pairs, such as 18-crown-6 and NO_3^-, 1,10-diaza-18-crown-6, and NO_3^- [26], and a single helical N_6 ligand (Figure 1, L′) and NO_3^- [21], respectively. In all these complexes, the central Ln(III) ion is surrounded by an (aza)crown ether or a helical ligand L′ in an equatorial manner, and axial positions are occupied by two or three nitrate anions; this leads to an easy-axial magnetic anisotropy and slow magnetic relaxation phenomena of Ce(III) and Nd(III) as Kramers ions. The radii of the macrocyclic and helical ligands are slightly larger than the ionic radii of the lanthanide ions, and hence the ligands show distortion [20,26] or helication [27] in the complex formation, hence the coordination distances are rather long and the coordination itself is weak. The mismatching between ligand radius and lanthanide radius becomes larger for heavy lanthanide ions and, for the case of macrocyclic ligands, we have not succeeded in synthesizing heavy lanthanide complexes with macrocyclic ligands to reveal the correlation between slow magnetic relaxation phenomena and the molecular structure accompanied with a series of axial ligands. In this study, we decided to introduce a smaller N_5 ligand L as an equatorial ligand, shown in Figure 1. The molecular radius of L is smaller than those of the macrocyclic ligands and the helical ligand L′. Complexes of L with heavy lanthanide ions, such as Tb(III) or Dy(III), will be sufficiently stable to exchange axial ligands with different donating abilities. Moreover, it is expected that L can occupy equatorial positions in a flatter manner than L′, being free from helication. In this study, we synthesized a family of Dy(III) complexes with the general formula of $[DyA_2L]^+$ (A^- = bidentate anionic *O*-donor ligand), such as $[Dy(NO_3)_2L]NO_3$ (**1**), $[Dy(AcO)_2L]CF_3SO_3$ (**2**), and $[Dy(acac)_2L]CF_3SO_3$ (**3**), to investigate the correlation between the axial ligand nature and slow magnetic relaxation phenomena. Syntheses, crystal structures, magnetic properties, and DFT calculation results of these complexes are discussed.

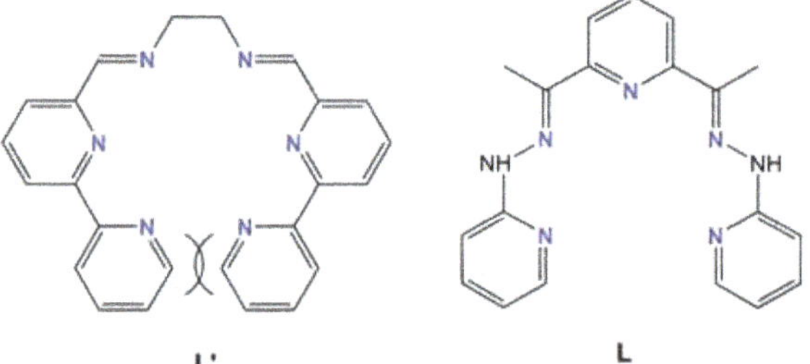

Figure 1. Structure of N_6 ligand L′ and N_5 ligand L.

2. Results

2.1. Synthesis and Characterization

We have newly synthesized three Dy(III) complexes by the reaction of an appropriate dysprosium salt with L in an organic solvent, such as $Dy(NO_3)_3 \cdot 5H_2O$ in MeCN for **1**, $Dy(AcO)_3 \cdot 4H_2O$ and $Dy(CF_3SO_3)_3$ in a 2:1 molar ratio in EtOH for **2**, and $Dy(acac)_3$ and $Dy(CF_3SO_3)_3$ in a 2:1 molar ratio in 2-propanol for **3**, to obtain crystals suitable for single X-ray crystallography. Their structures were revealed from the crystallographic analyses both for single crystals and microcrystalline samples (Powder X-ray diffraction (PXRD) data are given as Figure S1, in the Supplementary Information).

Crystal structures of the cationic part of three complexes are shown in Figure 2 and Figures S2–S5. Crystallographic data, accompanied by the selected distances and angles, are summarized in Tables S1 and S2. **1** and **2** crystallized in a triclinic crystal system with a *P*-1 space group, whereas **3** crystallized in monoclinic $P2_1/n$. In the complexes, a neutral L ligates as a pentadentate ligand which occupies equatorial positions forming four pentagonal chelate rings. Two anionic ligands, such as NO_3^- in **1**, AcO^- in **2**, or $acac^-$ in **3**, occupy the positions above and below the lanthanide ion in a bidentate fashion, to complete the N_5O_4 nona-coordination of the Dy(III) ion. In all cases, one whole molecule is crystallographically independent; however, each has pseudo two-fold symmetry along the axis passing through Dy and N4 of the central pyridine ring. L shows twist distortion around the pseudo two-fold axis, which can be evaluated from the values of the dihedral angles between the two terminal pyridine rings involving N1 and N7. The estimated angles were 25.45(11)°, 19.92(13)°, and 57.40(11)°, for **1**, **2**, and **3**, respectively, and strongly correlated with the axial ligands. The nitrate and acetate ligands coordinated with the formation of four-membered chelate rings and they were less bulky than acetylacetonate, which had coordinated forming six-membered chelate rings. In all complexes, the shortest contacts between O-donor and N-donor atoms (i.e., O2···N1 and O4···N7) were very similar within the range of 2.692(3) to 2.837(2) Å, indicating the presence of van der Waals contacts. The steric repulsion among axial nitrate or acetate ligands and equatorial L was small, and **1** and **2** had similar small values of dihedral angles. The steric repulsion in **3** is expected to be larger leading to a larger twisting of L. The twisting of L is also defined by the planarity of the five equatorially coordinating N-donors. The maximum deviations of the N-donor from the ideal planes defined by N_5Dy were estimated as 0.3534(11) Å, 0.3330(10) Å, and 0.6594(19) Å, respectively. The coordination distances in the complexes show a systematic difference in the series of **1**, **2**, and **3**. Dy–N distances elongated along this order: Dy–N distances were 2.4184(17)–2.4939(18) Å for **1**, 2.4547(16)–2.5018(18) Å for **2**, and 2.536(2)–2.696(3) Å for **3**. On the contrary, Dy–O distances shortened in this order: 2.3903(16)–2.4657(16) Å for **1**, 2.3923(16)–2.4357(16) Å for **2**, and 2.287(2)–2.370(2) Å for **3**. The longest Dy–N distances in **3** resulted from the twisting of L. The coordination distances of O-donors and N-donors in **1** were almost similar, whereas in **2**, Dy–O distances were slightly shorter than those of Dy–N. The molecular structures of **1** and **2** were very similar, with twisting of L, and the difference in coordination distances reflects the difference in negative charge distributions between nitrate and acetate O-donor atoms. Because a negative charge in the nitrate delocalized among three oxygen atoms, the charge density of O-donor atoms in the nitrate is slightly smaller than those of the acetate anion. Larger negative charges in the O-donor enforced a slight shortening of the coordination distance in **2**, and it also enforced a slight elongation of the Dy–N distances. The anisotropic ligand field is defined by both the difference in coordination distances and the difference in negative charge distributions of donor atoms. The overall ligand field geometries of **1**–**3** were similar, and could lead to an axially stressed ligand field by sandwiching the Dy(III) ion between negatively charged O-donor ligands. The observed structural trend suggests that the magnetic anisotropy is enhanced in the order of **1**, **2**, and **3** because the difference in the coordination distances become larger in this order. The larger negative charges and the shorter coordination distances of O-donors lead to a larger stress along z-axis, and this is likely to enhance the easy-axial magnetic anisotropy for Dy(III) as an oblate ion.

In **2**, crystalline solvent molecules showed heavy disordering. They were assigned as two EtOH and one H_2O, with 50%, 25%, and 50% occupancies according to the elemental analyses. These solvent molecules were closely located to O3 (~2.85 Å) suggesting the hydrogen bond formation, however, π electrons of the acetate ligand are symmetrically delocalized (O3-C22 = 1.263(3) Å and O4-C22 = 1.265(3) Å), and hence the negative charge distributions of O3 and O4 would be similar. Therefore we assumed that the influence of the hydrogen bond formations on the ligand field anisotropy around Dy(III) ion would be negligible.

The shortest Dy···Dy distances were estimated as 7.2871(13) Å (symmetry code: 1 - x, 1 - y, 1- z), 7.9634(10) Å (symmetry code: 1 - x, -y, 1 - z), and 9.1817(8) Å (symmetry code: 1 - x, 1 - y, -z). The Dy···Dy separations were long enough to avoid through-space magnetic interactions. The packing diagrams in the unit cell are given as Figures S2–S4.

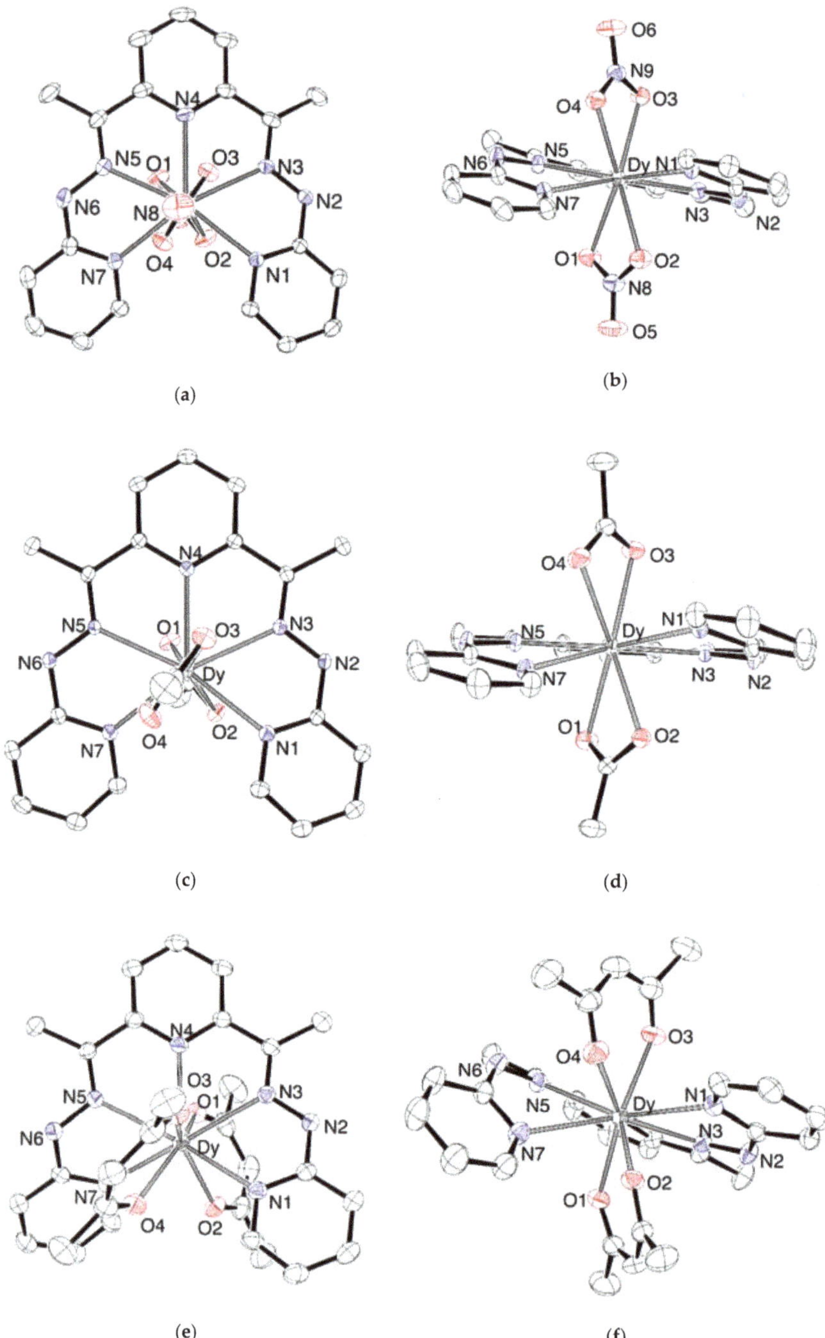

Figure 2. ORTEP drawing of the cationic parts of **1** (**top**), **2** (**central**), and **3** (**bottom**) at the 50% probability level. Hydrogen atoms are omitted for clarity: (**a,c,e**) Top view and (**b,d,f**) side view of the molecules.

2.2. Magnetic Properties of the Complexes

2.2.1. DC Susceptibility of the Complexes

DC susceptibility data were recorded at various temperatures for all complexes under an application of 1000 Oe DC field. Figure 3 shows the temperature dependence of $\chi_M T$ products. The Curie constant for Dy(III) ion in an isotropic ligand field was estimated as 14.2 emu K mol^{-1} ($J = 15/2$, g = 4/3), and the observed values at 300 K, 13.8, 16.0, and 14.6 emu K mol^{-1}, respectively, were close to the expected values. This may indicate that the magnetic anisotropy is small in these complexes. The $\chi_M T$ values were almost constant down to 150 K (for **1**) or 100 K (for **2** and **3**), and then $\chi_M T$ gradually decreased as the temperature was cooled, due to the thermal depopulation among the substates arising from the $^6H_{15/2}$ ground state which split under the anisotropic crystal field.

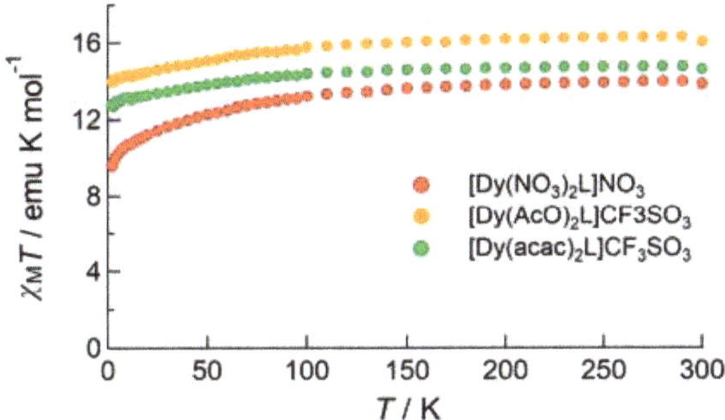

Figure 3. Temperature dependence of the $\chi_M T$ product for each complex measured under a 1000 Oe DC field applied condition.

2.2.2. AC Susceptibility of the Complexes

The slow magnetic relaxation of the complexes was revealed by measuring the alternating current (AC) magnetic susceptibility under the applied conditions of zero field or direct current (DC) field. For all complexes, slow magnetic relaxation phenomena were initially analyzed under a zero DC bias field (Figure S6), which exhibited no out-of-phase signals χ_M'' for **1**, and weak signals for **2** and **3** owing to the quantum-tunneling magnetization (QTM) relaxation process, which was faster for **1** and **3** than the flipping of the magnetic field. For **2**, χ_M'' shows a peak at ~6000 Hz, which is almost constant in the temperature range up to 6.0 K. This indicates that the magnetization relaxation occurs via QTM in this temperature range. The frequency dependences of χ_M'' were analyzed on the basis of the Cole–Cole equation (see below), and the QTM relaxation rate τ_{QTM} was estimated as 2.70(9) $\times$ 10^{-5} s. Upon applying the DC bias field, fast relaxation via the QTM process was suppressed and slow magnetic relaxations were observed for all complexes. To reveal the effects of the bias field on slow magnetic relaxations, the bias field dependences of the out-of-phase susceptibilities were measured under the DC field in the ranges of 0 to 3000 Oe and 0 to 5000 Oe for several temperatures (Figure 4).

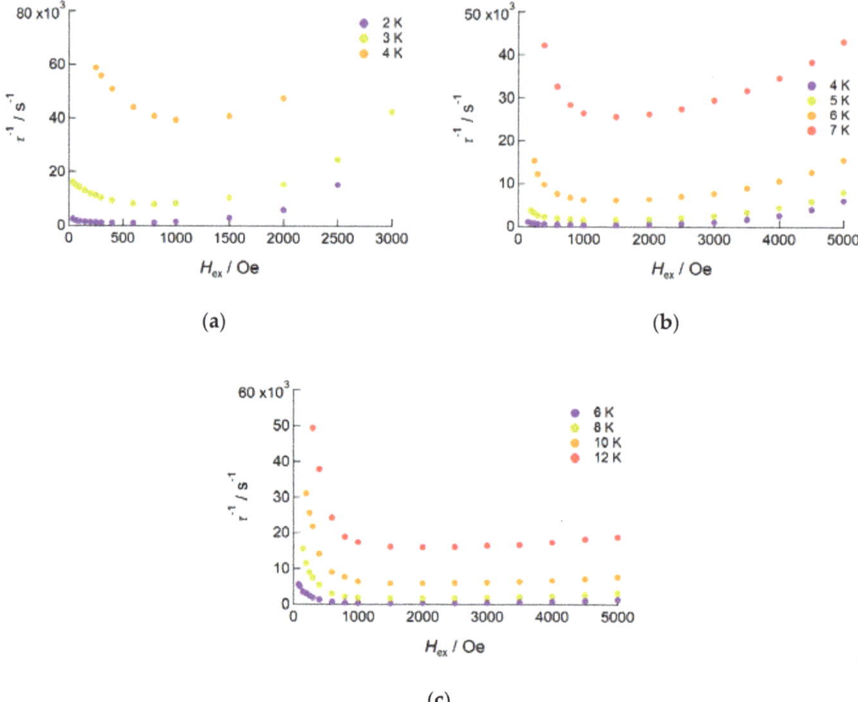

Figure 4. DC field dependence of the relaxation rate τ^{-1} of (**a**) **1**, (**b**) **2**, and (**c**) **3** measured in the temperature ranges of 2 to 4 K, 4 to 7 K, and 6 to 12 K, respectively.

In all complexes, slow magnetic relaxation occurred when a weak DC field of 200 Oe was applied, and it became slower as the applied field increased. The relaxation rate reached a minimum under a DC field of 1000 to 2000 Oe, and was enhanced again when the field was further strengthened, mainly due to the direct process of relaxation becoming predominant. Hence, the dynamic magnetic property of each complex was revealed under the application of 1000 Oe DC, where the QTM and direct processes were effectively suppressed. Under this condition, all complexes exhibited frequency-dependent in-phase ($\chi_M{}'$) and out-of-phase ($\chi_M{}''$) susceptibilities at temperatures up to 4.2 K, 7.8 K, and 14 K, for **1**, **2**, and **3** respectively, with an AC frequency up to 10,000 Hz. Figure 5 shows the frequency dependence of the $\chi_M{}'T$ products and $\chi_M{}''$ values.

1 and **2** showed bilaterally symmetric shapes of out-of-phase signals, and hence the AC susceptibility data were analyzed with the Cole–Cole Equation (1) given below [28].

$$\chi*(\omega) = \chi_S + \frac{\chi_T - \chi_S}{1 + (i\omega\tau)^{1-\alpha}} \tag{1}$$

Here χ_T and χ_S denote the isothermal and the adiabatic susceptibilities, respectively, τ denotes the relaxation time at each temperature, and α denotes the distribution of τ (Tables S3 and S4). At first, the frequency dependence of $\chi_M{}'T$ products was fitted using these four parameters, and the frequency dependences of $\chi_M{}''$ were reproduced using the estimated parameters, which showed good agreement with the observations. The estimated α values were small enough, and hence the slow magnetic relaxations occurred via a single process.

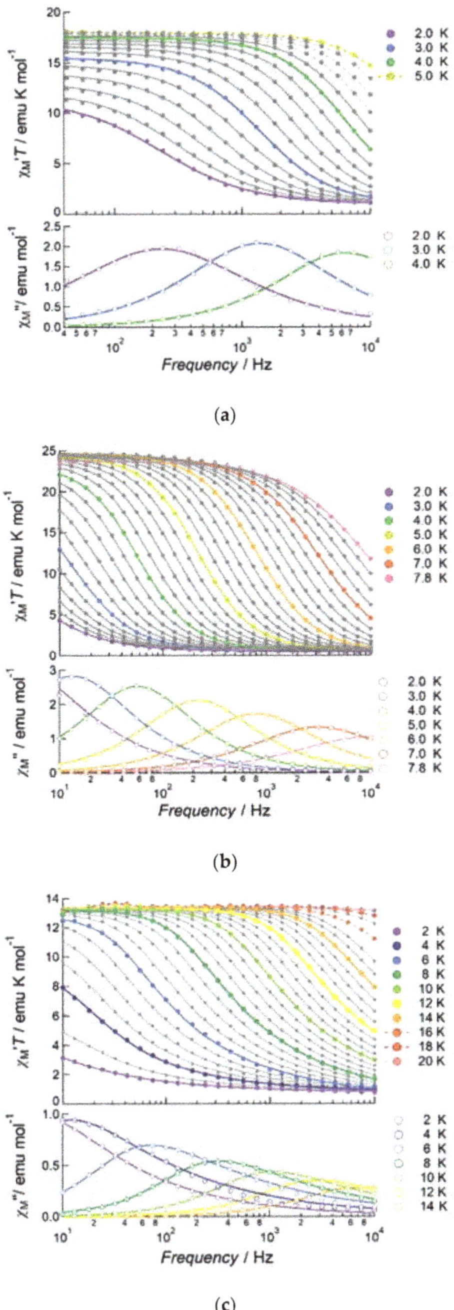

(a)

(b)

(c)

Figure 5. AC susceptibility data of (**a**) **1**, (**b**) **2**, and (**c**) **3** measured under an application of 1000 Oe bias field: frequency dependence of the products of temperature and in-phase susceptibility (closed circles) and out-of-phase susceptibility (open circles) measured at several temperatures. Solid curves represent theoretical calculations on the basis of the Cole–Cole equation for **1** and **2**, or the Cole–Davidson equation for **3**, of which the estimated parameters are listed in Tables S3–S5.

3 showed slightly asymmetric signals both for $\chi_M'T$ and χ_M'', which obeyed the Cole–Davidson Equation (2) given below [29].

$$\chi * (\omega) = \chi_S + \frac{\chi_T - \chi_S}{(1 + i\omega\tau)^{1-\beta}} \tag{2}$$

Here β denotes the distribution of τ. The analyses were carried out similarly to those for **1** and **2**; the frequency dependences of $\chi_M'T$ products were fitted with Equation (2) first, of which the estimated parameters were listed in Table S5, and then the frequency dependence of χ_M'' were reproduced. The β values were large at low temperature, with the largest value of 0.53 at 3.0–4.0 K; these became smaller at higher temperatures, reaching down to 0.41 at 13 K. Both α and β parameters describe the distributions of τ from the ideal value; however, their scales were slightly different. Figure 6 shows the theoretical calculations of χ'' for the different α and β values. In both cases, $\alpha = 0$ and $\beta = 0$ mean that the distribution of relaxation time is zero and all molecules flip with a unique relaxation time. This obeys the Debye equation. The peaks of both plots were normalized as 1.0 at $\alpha = 0$ and $\beta = 0$. When the distribution of τ becomes wider, the number of molecules which flip in the different relaxation time increases, and hence the peak height becomes lower depending on α and β. For the same height of the χ'' plots, β takes a value 2.0–2.8 times larger than that of α. Considering this relationship, the β values for **3** correspond to the α values at around 0.2, and hence the distribution of τ is regarded to be sufficiently small. Hence, we assumed **3** to be a field-induced SMM.

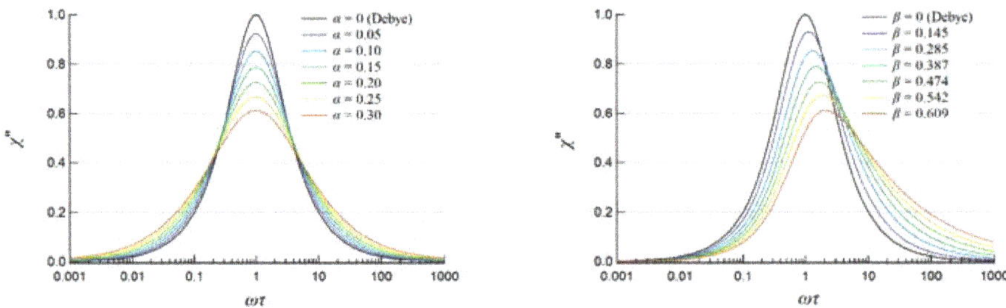

Figure 6. Theoretical curves of the out-of-phase susceptibility χ'' as a function of $\omega\tau$ for several α values (**left**) or β values (**right**), calculated on the basis of the Cole–Cole and Cole–Davidson equations. At $\alpha = 0$ and $\beta = 0$, both plots obey Debye equation. The value of χ'' was normalized as 1.0 at $\omega\tau = 0$ and $\alpha = 0$ for the left plot and at $\omega\tau = 0$ and $= 0$ for the right plot.

Next we examined the temperature dependence of relaxation time τ using two methods. The left column in Figure 7 shows the Arrhenius plots for three complexes measured under a 1000 Oe DC applied field. Bent plots were obtained for all complexes over the entire temperature range. The QTM and direct processes were effectively suppressed under these conditions (Figure 4), and hence the bent shaped plots were due to the presence of Raman and Orbach processes; these two processes dominate in different temperature ranges. The data were first analyzed using the linear Arrhenius equation for the data in high temperature regions, where the Orbach process is predominant. This gave the best fit parameters of $\Delta E/k_B = 19.1(3)$ K and $\tau_0^{-1} = 2.1(2) \times 10^{-7}$ s for **1**, 82(3) K and $5(2) \times 10^{-10}$ s for **2**, and 84(3) K and $5(1) \times 10^{-8}$ s for **3**.

The right column in Figure 7 shows the temperature dependence of τ in double logarithm plots. It is known that the Raman process dominates in low temperature regions and that it obeys Equation (3) below. In plots of $\ln(\tau)$ vs. $\ln(T)$, relaxation via the Raman process is easily identified as a linear region found at low temperatures. The data here were analyzed using Equation (3), which gave the best fit parameters of $C = 96(15)$ s^{-1} K$^{-4.0}$ and $n = 4.0(2)$ for **1**, $C = 0.17(3)$ s^{-1} K$^{-5.42}$ and $n = 5.42(13)$ for **2**, and $C = 0.048(4)$ s^{-1} K$^{-5.12}$ and $n = 5.12(4)$ for **3**.

$$\tau^{-1} = CT^n, \quad \ln\tau = -\ln C - n\ln T \tag{3}$$

Finally, the whole dataset was analyzed using Equation (4) which considers both Raman and Orbach processes [30,31].

$$\tau^{-1} = CT^n + \tau_0^{-1} \exp\!\left(\frac{-\Delta E}{k_B T}\right),\tag{4}$$

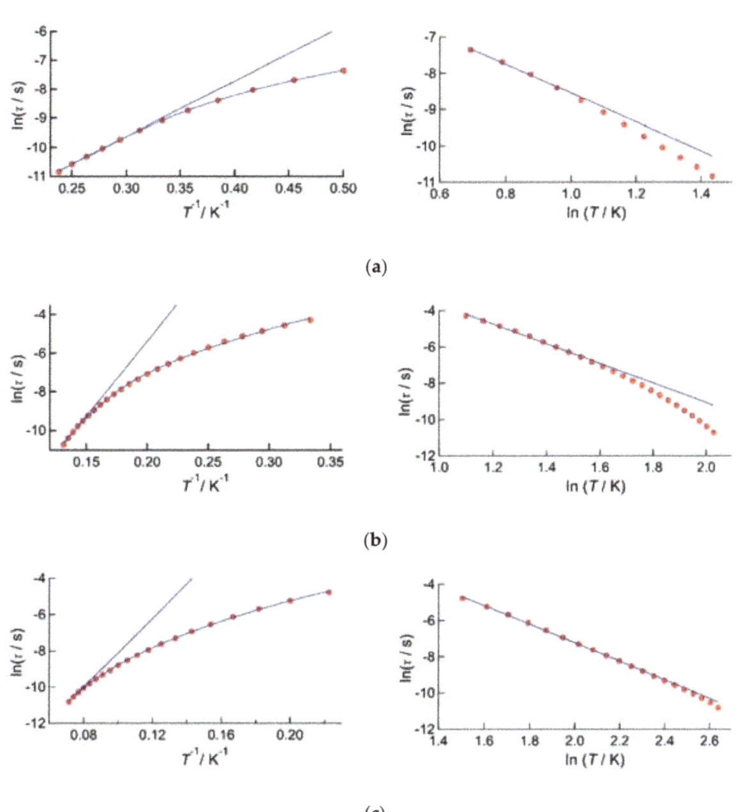

Figure 7. Temperature dependence of the flipping time of magnetization τ measured for (**a**) **1**, (**b**) **2**, and (**c**) **3** under an application of 1000 Oe DC field. (**left**) Arrhenius plots and (**right**) $\ln(\tau)$ vs. $\ln(T)$ plots. The blue lines represent theoretical calculations based on the linear Arrhenius equation for the Orbach process (**left**) or based on Equation (3) for the Raman process (**right**). The blue curves represent theoretical calculations considering both Orbach and Raman processes, expressed as Equation (4).

The estimated values are summarized in Table 1. Some parameters differ from the values based on the simple Raman or Orbach process. This may be due to the narrow range of temperatures where slow magnetic relaxation was observed.

Table 1. Best fit kinetic parameters for magnetization flipping estimated for complexes **1**–**3** on the basis of the Equation (4).

Complex	1	2	3
$\Delta E\, k_B^{-1}/K$	24.1(6)	85(3)	140(15)
τ_0/s	$9.0(9) \times 10^{-8}$	$5(2) \times 10^{-10}$	$3(3) \times 10^{-9}$
n	3.09(15)	5.4(1)	5.06(4)
$C/s^{-1}\, K^{-n}$	174(18)	0.17(2)	0.054(4)

From the structural analyses, the crystal field anisotropy as well as the magnetic anisotropy is enhanced in the order of **1** to **3**, and the kinetic parameters in Table 1 clearly support the idea. This result indicates that magnetic relaxation phenomena are strongly dependent on the nature of axial ligands, and that the increase of negative charge distributions on *O*-donor atoms plays a primitive role in enhancing the magnetic anisotropy. However, the structural changes from complex **1** with NO_3^- axial ligands to **2** with AcO^- axial ligands were small, while the increase of $\Delta E/k_B$ values was rather drastic. This may suggest the presence of other factors enhancing the barrier height of the magnetization flipping; hence we examined the electronic structures of the complexes with the DFT technique shown below.

2.3. Molecular Orbital Analyses

To elucidate the electronic structures of these complexes, we performed density functional theory (DFT) calculations for three model structures **m1–m3**, pertaining to complexes **1–3**, respectively, as illustrated in Figure 8. Calculated atomic spin densities of the Dy(III) ions in each model summarized in Table 2 are confirmed to be ~5.0, showing that the appropriate electronic configurations are obtained. Kohn-Sham orbital shapes and the energies of *f* orbitals are summarized in Figure S7 in the Supplementary Information, together with orbital energies of the highest occupied molecular orbital (HOMO). The energy level of occupied *f* electrons is approximately −10 to −13 eV, much lower than those of HOMO; this suggests that those are independent of the effects of frontier orbitals. On the other hand, those energy levels are comparable to the levels of coordinating σ orbitals and the stable π orbitals of ligands. As a consequence, some *f* orbitals can be hybridized with those ligand orbitals, although it is usually assumed that the *f* orbitals do not interact with ligand orbitals. As depicted in Figure S7, the *f* orbitals of **m1** are almost localized, except for HOMO-39(a) and HOMO-46(a) that are also found in equatorial ligand, while those of **m2** and **m3** are more delocalized by hybridization especially with σ and π orbitals of axial AcO^- and $acac^-$ ligands, respectively. In addition, the atomic spin densities of the Dy(III) ions in **m2** and **m3** are slightly smaller than those in **m1**. The result also indicates that the *f* spins are slightly delocalized to the ligands in **m2** and **m3**. Hence, the *f* electrons are considered to interact with axial ligand orbitals in **m2** and **m3** but not in **m1**, suggesting that the magnetic anisotropy of **m2** and **m3** can be affected by such hybridization between *f* and ligand orbitals.

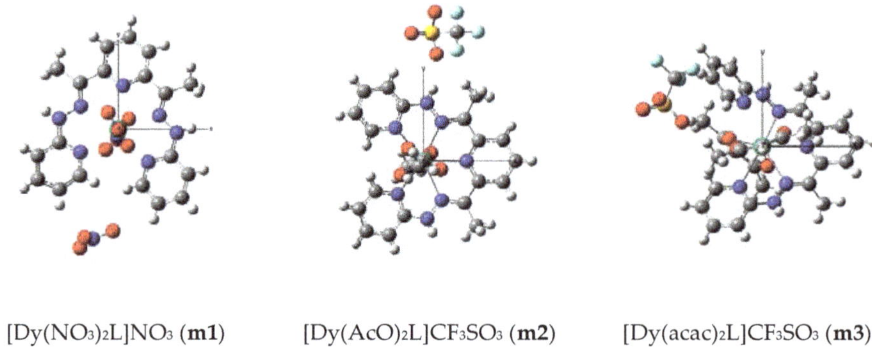

[Dy(NO₃)₂L]NO₃ (**m1**)   [Dy(AcO)₂L]CF₃SO₃ (**m2**)   [Dy(acac)₂L]CF₃SO₃ (**m3**)

Figure 8. Model structures **m1–m3** for complexes **1–3**, respectively.

Table 2. Calculated atomic spin densities assigned to each f orbitals.

Orbitals	m1	m2	m3
f_0	0.49	0.43	0.80
f_{+1}	0.98	0.72	0.64
f_{-1}	0.88	0.84	0.35
f_{+2}	0.51	0.86	0.73
f_{-2}	0.55	0.21	0.72
f_{+3}	0.61	0.85	0.74
f_{-3}	0.93	0.51	0.44
Sum of f orbitals	4.96	4.42	4.41

3. Materials and Methods

3.1. General Procedures and Methods

All chemicals and reagents were of reagent grade and used without further purification. All chemical reactions and sample preparations for physical measurements were performed in an ambient atmosphere. Variable temperature magnetic susceptibility measurements were performed on PPMS-9 and MPMS-XL magnetometers (Quantum Design). Diamagnetic corrections for each sample were applied using Pascal's constants. Elemental analyses were carried out with the help of the Research and Analytical Centre for Giant Molecules, Graduate School of Science, Tohoku University.

3.2. Synthesis of Complexes

Synthesis of L: The ligand L was prepared according to a previously reported method [32]. 10 mmol of 2,6-diacetylpyridine and 20 mmol of 2-pyridylhydrazine were dissolved into 25 mL of EtOH and one drop of conc. HCl solution was added. The resulting solution was stirred at 60 °C for 1 h to give a pale-yellow precipitate. The solution was left to stand for one night at room temperature to complete the reaction, and then the precipitate was filtrated under reduced pressure, washed with EtOH, and dried in vacuo. Yield 3.01 g (8.7 mmol, 87%).

Synthesis of $[\text{DyL(NO}_3)_2]\text{NO}_3$ (**1**): To the suspension of L (34.3 mg, 0.1 mmol) in MeCN (5 mL), 5 mL of 0.02 M $\text{Dy(NO}_3)_3\cdot5\text{H}_2\text{O}$ solution in MeCN was added. The resulting yellow-orange solution was left to stand for several days to give yellow prismatic crystals of **1** suitable for X-ray crystallography. Yield 52 mg, 75%. Elemental Anal. Calcd. for $[\text{Dy(L)(NO}_3)_2]\text{NO}_3$ ($\text{C}_{19}\text{H}_{19}\text{DyN}_{10}\text{O}_9$) C, 32.89; H, 2.76; N, 20.19. Found C, 32.92; H 2.78; N, 20.07.

Synthesis of $[\text{DyL(AcO)}_2]\text{CF}_3\text{SO}_3$ (**2**): With the suspension of L (20.7 mg, 0.06 mmol) in EtOH (2 mL), 0.8 mL of 0.05 M ethanolic solution of $\text{Dy(AcO)}_3\cdot4\text{H}_2\text{O}$ and 0.2 mL of 0.1 M ethanolic solution of $\text{Dy(CF}_3\text{SO}_3)_3$ were reacted. The resulting yellow-orange solution was sealed and left to stand for several days, giving yellow cubic crystals of **2** suitable for X-ray crystallography. Yield 35 mg, 72%. Elemental Anal. Calcd. for $[\text{Dy(L)(AcO)}_2]\text{CF}_3\text{SO}_3\cdot0.75\text{EtOH}\cdot0.5\text{H}_2\text{O}$ ($\text{C}_{25.5}\text{H}_{30.5}\text{DyF}_3\text{N}_7\text{O}_{8.25}\text{S}$) C, 37.41; H, 3.76; N, 11.98. Found C, 37.49; H 3.66; N, 12.15.

Synthesis of $[\text{DyL(acac)}_2]\text{CF}_3\text{SO}_3$ (**3**): To the suspension of L (20.6 mg, 0.06 mmol) in 2-propanol (2 mL), 0.8 mL of 0.05 M Dy(acac)_3 and 0.2 mL of 0.1 M $\text{Dy(CF}_3\text{SO}_3)_3$ solutions in the same solvent were added. The resulting yellow-orange solution was sealed and left to stand for several days to give yellow prismatic crystals of **3** suitable for X-ray crystallography. Yield 15 mg, 29%. Elemental Anal. Calcd. for $[\text{Dy(L)(acac)}_2]\text{CF}_3\text{SO}_3\cdot\text{H}_2\text{O}$ ($\text{C}_{30}\text{H}_{35}\text{DyF}_3\text{N}_7\text{O}_8\text{S}$) C, 41.26; H, 4.04; N, 11.23. Found C, 41.16; H 3.95; N, 11.24.

3.3. Crystallography

A single crystal of each complex was mounted on a Rigaku Varimax Saturn area detector for data collection using confocal monochromated MoKα radiation at low temperature (153 K). Intensity data were corrected for absorption using an empirical method included in the Crystal Clear software [33].

The structures were solved by direct methods with SIR-97 [34], and structure refinement was carried out using the full-matrix least-squares method on SHELXL-2013 [35]. Non-hydrogen atoms were anisotropically refined and hydrogen atoms were treated using the riding model. Crystallographic data are summarized in Tables S1 and S2 of the Supplementary Information. The complete crystal structure results, including bond lengths, angles, and atomic coordinates, are available as a CIF file in the Supplementary Information. The CCDC numbers are 1903775, 1903776, and 1903777 for compounds **1**, **2**, and **3**, respectively.

3.4. DFT Calculation

DFT calculations were performed for the model structures of **m1**–**m3**. Cartesian coordinates of the models summarized in Table S7 in the Supplementary Information are taken from the results of the above X-ray crystallographic analyses. All calculations were carried out using a Becke 3-parameter Lee-Yang-Parr hybrid functional set (B3LYP) [36] with a spin-unrestricted method under the gas phase condition. As a basis set, the Stuttgart RSC 1997 effective core potential [37] was used for the Dy atom in all models. 6-31+G* and 6-31G* were used for NO_3^- in **m1** and the other ligands, respectively. All calculations were performed using Gaussian09 [38].

4. Conclusions

Using N_5 ligand L, three Dy(III) complexes with similar coordination structures were synthesized. In each complex, equatorial positions of Dy(III) ion were occupied by five N-donor atoms from L, and the up- and down-sides were occupied by anionic O-donor atoms aiming to achieve the easy-axis magnetic anisotropy of an oblate type Dy(III) ion. In an anisotropic crystal field, the oblate shaped electronic cloud of the Kramers pair with the most Ising character was relatively stable. The complexes showed slow magnetic relaxation phenomena under application of a 1000 Oe bias DC field, and the barrier heights for the magnetization flipping were estimated as $\Delta E/k_B$ = 24.1(6) K, 85(3) K, and 140(15) K from the Arrhenius analyses, considering both Orbach and Raman relaxation processes. This order of barrier height is consistent with the strength of the crystal field anisotropy assessed from structural analysis of characteristics such as the Dy-O and Dy-N distances. However, the difference of the $\Delta E/k_B$ values between **1** and **2** were unexpectedly greater than the difference in coordinating structures. From the DFT calculations, it was found that the π character of the axial ligand plays significant role in the enhancement of magnetic anisotropy. In **2** and **3**, the results indicated that the interactions between f orbitals of Dy(III) and both σ and π orbitals of AcO$^-$ and acac$^-$ ligands were small but not ignorable, and that this may cause the presence of stronger magnetic anisotropy than in **1**. Our calculation results are not quantitative at present, and the prediction requires experimental confirmation; however, the idea could give a new perspective in designing SMMs with lanthanide(III) ions. Such an investigation is in progress in our group.

Supplementary Materials: The following are available online at http://www.mdpi.com/2312-7481/5/2/27/s1, Figure S1: PXRD patterns of **1**–**3**, Figures S2–S4: crystal packing of **1**–**3**, Figure S5: Ortep drawings of **1**–**3**, Figure S6: frequency dependence of $\chi_M'T$ and χ_M'' of **1**–**3** measured under zero bias field condition, Figure S7: Kohn–Sham orbitals of f orbitals for models **m1**–**m3**, Table S1: crystallographic data for **1**–**3**, Table S2: selected bond distances and angles for **1**–**3**, Table S3: best fitted parameters using Cole–Cole equation for **1**, Table S4: best fitted parameters using Cole–Cole equation for **2**, Table S5: best fitted parameters using Cole–Davidson equation for **3**, Table S6: Cartesian coordinates of **m1**–**m3** for DFT calculations.

Author Contributions: K.K. and Y.H. prepared and characterized the complexes. K.K. and T.K. performed the magnetic measurements and analyzed the data. K.I., M.N., and Y.K. performed the DFT calculations and analyzed the results. All the authors reviewed the paper.

Funding: This work was supported in part by grant-in-aid for Scientific Research (C) (No. 17K05811) from JSPS, Japan (T.K.)

Conflicts of Interest: The authors declare no conflicts of interest.

Abbreviations

ORTEP   Oak Ridge Thermal-Ellipsoid Plot program

References

1. Gatteschi, D.; Sessoli, R.; Villain, J. *Molecular Nanomagnets*; Oxford University Press: New York, NY, USA, 2006.
2. Layfield, R.A. Organometallic Single-Molecule Magnets. *Organometallics* **2014**, *33*, 1084–1099. [CrossRef]
3. Madhu, N.T.; Tang, J.-K.; Hewitt, I.J.; Clérac, R.; Wernsdorfer, W.; van Slageren, J.; Anson, C.E.; Powell, A.K. What makes a single molecule magnet? *Polyhedron* **2005**, *24*, 2864–2869. [CrossRef]
4. Glaser, T. Rational design of single-molecule magnets: A supramolecular approach. *Chem. Commun.* **2011**, *47*, 116–130. [CrossRef]
5. Woodruff, D.N.; Winpenny, R.E.; Layfield, R.A. Lanthanide single-molecule magnets. *Chem. Rev.* **2013**, *113*, 5110–5148. [CrossRef] [PubMed]
6. Zhang, P.; Guo, Y.-N.; Tang, J. Recent advances in dysprosium-based single molecule magnets: Structural overview and synthetic strategies. *Coord. Chem. Rev.* **2013**, *257*, 1728–1763. [CrossRef]
7. Feltham, H.L.C.; Brooker, S. Review of purely 4f and mixed-metal nd-4f single-molecule magnets containing only one lanthanide ion. *Coord. Chem. Rev.* **2014**, *276*, 1–33. [CrossRef]
8. Liddle, S.T.; van Slageren, J. Improving f-element single molecule magnets. *Chem. Soc. Rev.* **2015**, *44*, 6655–6669. [CrossRef]
9. Meng, Y.S.; Jiang, S.D.; Wang, B.W.; Gao, S. Understanding the Magnetic Anisotropy toward Single-Ion Magnets. *Acc. Chem. Res.* **2016**, *49*, 2381–2389. [CrossRef] [PubMed]
10. Day, B.M.; Guo, F.S.; Layfield, R.A. Cyclopentadienyl Ligands in Lanthanide Single-Molecule Magnets: One Ring to Rule Them All? *Acc. Chem. Res.* **2018**, *51*, 1880–1889. [CrossRef] [PubMed]
11. Chen, Y.C.; Liu, J.L.; Ungur, L.; Liu, J.; Li, Q.W.; Wang, L.F.; Ni, Z.P.; Chibotaru, L.F.; Chen, X.M.; Tong, M.L. Symmetry-Supported Magnetic Blocking at 20 K in Pentagonal Bipyramidal Dy(III) Single-Ion Magnets. *J. Am. Chem. Soc.* **2016**, *138*, 2829–2837. [CrossRef]
12. Goodwin, C.A.P.; Ortu, F.; Reta, D.; Chilton, N.F.; Mills, D.P. Molecular magnetic hysteresis at 60 kelvin in dysprosocenium. *Nature* **2017**, *548*, 439–442. [CrossRef]
13. Yu, S.; Hu, Z.; Chen, Z.; Li, B.; Zhang, Y.Q.; Liang, Y.; Liu, D.; Yao, D.; Liang, F. Two Dy(III) Single-Molecule Magnets with Their Performance Tuned by Schiff Base Ligands. *Inorg. Chem.* **2019**. [CrossRef] [PubMed]
14. Yang, J.W.; Tian, Y.M.; Tao, J.; Chen, P.; Li, H.F.; Zhang, Y.Q.; Yan, P.F.; Sun, W.B. Modulation of the Coordination Environment around the Magnetic Easy Axis Leads to Significant Magnetic Relaxations in a Series of 3d-4f Schiff Complexes. *Inorg. Chem.* **2018**, *57*, 8065–8077. [CrossRef] [PubMed]
15. Shen, F.X.; Li, H.Q.; Miao, H.; Shao, D.; Wei, X.Q.; Shi, L.; Zhang, Y.Q.; Wang, X.Y. Heterometallic M(II)Ln(III) (M = Co/Zn; Ln = Dy/Y) Complexes with Pentagonal Bipyramidal 3d Centers: Syntheses, Structures, and Magnetic Properties. *Inorg. Chem.* **2018**. [CrossRef] [PubMed]
16. Ehama, K.; Ohmichi, Y.; Sakamoto, S.; Fujinami, T.; Matsumoto, N.; Mochida, N.; Ishida, T.; Sunatsuki, Y.; Tsuchimoto, M.; Re, N. Synthesis, structure, luminescent, and magnetic properties of carbonato-bridged $Zn(II)_2Ln(III)_2$ complexes $[(m_4\text{-}CO_3)_2\{Zn(II)L(n)Ln(III)(NO_3)\}_2]$ (Ln(III) = Gd(III), Tb(III), Dy(III); L(1) = *N,N′*-bis(3-methoxy-2-oxybenzylidene)-1,3-propanediaminato, L(2) = *N,N′*-bis(3-ethoxy-2-oxybenzylidene)-1,3-propanediaminato). *Inorg. Chem.* **2013**, *52*, 12828–12841. [CrossRef] [PubMed]
17. Long, J.; Basalov, I.V.; Forosenko, N.V.; Lyssenko, K.A.; Mamontova, E.; Cherkasov, A.V.; Damjanovic, M.; Chibotaru, L.F.; Guari, Y.; Larionova, J.; et al. Dysprosium Single-Molecule Magnets with Bulky Schiff Base Ligands: Modification of the Slow Relaxation of the Magnetization by Substituent Change. *Chemistry* **2019**, *25*, 474–478. [CrossRef]
18. Bar, A.K.; Kalita, P.; Sutter, J.P.; Chandrasekhar, V. Pentagonal-Bipyramid Ln(III) Complexes Exhibiting Single-Ion-Magnet Behavior: A Rational Synthetic Approach for a Rigid Equatorial Plane. *Inorg. Chem.* **2018**, *57*, 2398–2401. [CrossRef] [PubMed]
19. Long, J.; Tolpygin, A.O.; Cherkasov, A.V.; Lyssenko, K.A.; Guari, Y.; Larionova, J.; Trifonov, A.A. Single-Molecule Magnet Behavior in Dy^{3+} Half-Sandwich Complexes Based on Ene-Diamido and Cp* Ligands. *Organometallics* **2019**, *38*, 748–752. [CrossRef]
20. Maxwell, L.; Amoza, M.; Ruiz, E. Mononuclear Lanthanide Complexes with 18-Crown-6 Ether: Synthesis, Characterization, Magnetic Properties, and Theoretical Studies. *Inorg. Chem.* **2018**, *57*, 13225–13234. [CrossRef]

21. Wada, H.; Ooka, S.; Iwasawa, D.; Hasegawa, M.; Kajiwara, T. Slow Magnetic Relaxation of Lanthanide(III) Complexes with a Helical Ligand. *Magnetochemistry* **2016**, *2*, 43. [CrossRef]

22. Ji, C.-L.; Jiang, Y.-X.; Zhang, J.-C.; Qi, Z.-Y.; Kong, J.-J.; Huang, X.-C. Field-induced Slow Magnetic Relaxation Behavior in a Mononuclear Dy(III) Complex based on 8-Hydroxyquinoline Derivate Ligand. *Zeitschrift für Anorganische und Allgemeine Chemie* **2018**, *644*, 1635–1640. [CrossRef]

23. Then, P.L.; Takehara, C.; Kataoka, Y.; Nakano, M.; Yamamura, T.; Kajiwara, T. Structural switching from paramagnetic to single-molecule magnet behaviour of $LnZn_2$ trinuclear complexes. *Dalton Trans.* **2015**, *44*, 18038–18048. [CrossRef] [PubMed]

24. Hino, S.; Maeda, M.; Yamashita, K.; Kataoka, Y.; Nakano, M.; Yamamura, T.; Nojiri, H.; Kofu, M.; Yamamuro, O.; Kajiwara, T. Linear trinuclear Zn(II)-Ce(III)-Zn(II) complex which behaves as a single-molecule magnet. *Dalton Trans.* **2013**, *42*, 2683–2686. [CrossRef] [PubMed]

25. Takehara, C.; Then, P.L.; Kataoka, Y.; Nakano, M.; Yamamura, T.; Kajiwara, T. Slow magnetic relaxation of light lanthanide-based linear $LnZn_2$ trinuclear complexes. *Dalton Trans.* **2015**, *44*, 18276–18283. [CrossRef] [PubMed]

26. Wada, H.; Ooka, S.; Yamamura, T.; Kajiwara, T. Light Lanthanide Complexes with Crown Ether and Its Aza Derivative Which Show Slow Magnetic Relaxation Behaviors. *Inorg. Chem.* **2017**, *56*, 147–155. [CrossRef]

27. Hasegawa, M.; Ohtsu, H.; Kodama, D.; Kasai, T.; Sakurai, S.; Ishii, A.; Suzuki, K. Luminescence behaviour in acetonitrile and in the solid state of a series of lanthanide complexes with a single helical ligand. *New J. Chem.* **2014**, *38*, 1225–1234. [CrossRef]

28. Cole, K.S.; Cole, R.H. Dispersion and Absorption in Dielectrics I. Alternating Current Characteristics. *J. Chem. Phys.* **1941**, *9*, 341–351. [CrossRef]

29. Davidson, D.W.; Cole, R.H. Dielectric Relaxation in Glycerol, Propylene Glycol, andn-Propanol. *J. Chem. Phys.* **1951**, *19*, 1484–1490. [CrossRef]

30. Abragam, A.; Bleaney, B. *Electron Paramagnetic Resonance of Transition Ions*; Oxford University Press: Oxford, UK, 1970.

31. Carlin, R.L. *Magnetochemistry*; Springer: Berlin/Heidelberg, Germany, 1986.

32. Sakamoto, M.; Matsumoto, N.; Okawa, H. Synthesis and Molecular Structure of Europium(III) Complex with 2,6-Diacetylpyridine Bis(2-pyridylhydrazone). *Bull. Chem. Soc. Jpn.* **1991**, *64*, 691–693. [CrossRef]

33. *Crystal Clear, Operating Software for the CCD Detector System*; Version 1.3.5; Rigaku and Molecular Structure Corp.: Tokyo, Japan; The Woodlands, TX, USA, 2003.

34. Altomare, A.; Burla, M.C.; Camalli, M.; Cascarano, G.L.; Giacovazzo, C.; Guagliardi, A.; Moliterni, A.G.G.; Polidori, G.; Spagna, R. SIR97: A new tool for crystal structure determination and refinement. *J. Appl. Crystallogr.* **1999**, *32*, 115–119. [CrossRef]

35. Sheldrick, G. Crystal structure refinement with SHELXL. *Acta Crystallogr. Sect. C Struct. Chem.* **2015**, *71*, 3–8. [CrossRef] [PubMed]

36. Becke, A. Density-functional thermochemistry. III. The role of exact exchange. *J. Chem. Phys.* **1993**, *98*, 5648–5652. [CrossRef]

37. Andrae, D.; Häußermann, U.; Dolg, M.; Stoll, H.; Preuß, H. Energy-adjustedab initio pseudopotentials for the second and third row transition elements. *Theor. Chim. Acta* **1990**, *77*, 123–141, *Note that this basis set is quasi-relativistic and sometimes called MWB28 or SDD*. [CrossRef]

38. Frisch, M.J.; Trucks, G.W.; Schlegel, H.B.; Scuseria, G.E.; Robb, M.A.; Cheeseman, J.R.; Scalmani, G.; Barone, V.; Mennucci, B.; Petersson, G.A.; et al. *Gaussian 09*; Revision C01; Gaussian, Inc.: Wallingford, CT, USA, 2009.

 *magnetochemistry*

Article

Series of Chloranilate-Bridged Dinuclear Lanthanide Complexes: Kramers Systems Showing Field-Induced Slow Magnetic Relaxation [†]

Ryuta Ishikawa [1,*], Shoichi Michiwaki [1], Takeshi Noda [1], Keiichi Katoh [2], Masahiro Yamashita [2,3,4] and Satoshi Kawata [1]

[1] Department of Chemistry, Faculty of Science, Fukuoka University, 8-19-1 Nanakuma, Jonan-ku, Fukuoka 814-0180, Japan; sd142022@cis.fukuoka-u.ac.jp (S.M.); sd172011@cis.fukuoka-u.ac.jp (T.N.); kawata@fukuoka-u.ac.jp (S.K.)

[2] Department of Chemistry, Graduate School of Science, Tohoku University, 6-3 Aza-Aoba, Aramaki, Aoba-ku, Sendai, Miyagi 980-8578, Japan; kkatoh@m.tohoku.ac.jp (K.K.); yamasita@agnus.chem.tohoku.ac.jp (M.Y.)

[3] Advanced Institute for Materials Research (AIMR), Tohoku University, 2-1-1 Katahira, Aoba-ku, Sendai, Miyagi 980-8577, Japan

[4] School of Materials Science and Engineering, Nankai University, Tianjin 300350, China

* Correspondence: ryutaishikawa@fukuoka-u.ac.jp; Tel.: +81-92-871-6631

† This manuscript is dedicated to Professor Masahiro Yamashita on the occasion of his 65th birthday for his contributions to the multifunctional nanosciences of advanced coordination compounds.

Received: 26 March 2019; Accepted: 18 April 2019; Published: 2 May 2019

Abstract: A series of chloralilate-bridged dinuclear lanthanide complexes of formula $[\{Ln^{III}(Tp)_2\}_2(\mu\text{-}Cl_2An)]\cdot2CH_2Cl_2$, where Cl_2An^{2-} and Tp^- represent the chloranilate and hydrotris (pyrazolyl)borate ligands, respectively, and Ln = Gd (**1**), Tb (**2**), Ho (**3**), Er (**4**), and Yb (**5**) was synthesized. All five complexes were characterized by an elemental analysis, infrared spectroscopy, single crystal X-ray diffraction, and SQUID measurements. The complexes **1–5** in the series were all isostructural. A comparison of the temperature dependence of the dc magnetic susceptibility data of these complexes revealed clear differences depending on the lanthanide center. Ac magnetic susceptibility measurements revealed that none of the five complexes exhibited a slow magnetic relaxation under a zero applied dc field. On the other hand, the Kramers systems (complexes **4** and **5**) clearly displayed a slow magnetic relaxation under applied dc fields, suggesting field-induced single-molecule magnets that occur through Orbach and Raman relaxation processes.

Keywords: lanthanide ions; slow magnetic relaxation; single-molecule magnets

1. Introduction

Single-molecule magnets (SMMs) and single-ion magnets (SIMs), comprising molecules with a one spin center, have attracted significant attention as potential candidates for molecule-based electronic applications such as high-density information storage [1], quantum computing [2–5], and spintronic devices [6–9]. For the realization of such applications, a very large barrier height and a high blocking temperature for the reorientation of the magnetic moment must be achieved. The barrier height and blocking temperature in SMMs and SIMs depend on two key factors, namely, a significant magnetic anisotropy and large number of spins. Thus, the high magnetic anisotropy and large number of spins per ion make lanthanide(III) ions (Ln^{III}) highly suitable for application in SMMs and SIMs. In fact, the development of SMMs and SIMs based on Ln complexes is on the rise [10–18]. These Ln^{III}-based SMMs and SIMs generally display very high barrier heights and blocking temperatures over those of the representative cluster-type SMM $[Mn^{III}_8Mn^{IV}_4O_{12}(CH_3COO)_{16}(H_2O)_{24}]\cdot2CH_3COOH\cdot4H_2O$, which is based on first-row transition metal ions [19,20].

Recently, chloralilate (Cl_2An^{2-}) bridged dinuclear Ln^{III} complexes of the formula $[\{Ln^{III}(Tp)_2\}_2(\mu\text{-}Cl_2An)]\cdot2CH_2Cl_2$ (Tp^- = hydrotris(pyrazolyl)borate) have been reported in the literature [21–24]. In these complexes, the Dy^{III} analogue displays a slow magnetic relaxation under small applied dc magnetic fields, thus behaving as a field-induced SMM [22–24]. This paper reports the syntheses, structures, and magnetic properties of a series of Cl_2An^{2-} bridged dinuclear Ln complexes of the formula $[\{Ln(Tp)_2\}_2(\mu\text{-}Cl_2An)]\cdot2CH_2Cl_2$ [Ln = Gd (**1**), Tb (**2**), Ho (**3**), Er (**4**), and Yb (**5**)] to systematically investigate the magnetic properties in other Ln^{III} analogues of $[\{Ln^{III}(Tp)_2\}_2(\mu\text{-}Cl_2An)]\cdot2CH_2Cl_2$.

2. Results and Discussions

2.1. Structural Descriptions

The series of Cl_2An^{2-} bridged neutral dinuclear Ln^{III} complexes **1–5** were prepared by a slight modification of the original method reported by Kaizaki et al. [21]; some of these complexes have been previously reported [22–24]. All the complexes were isolated as X-ray quality single crystals through several recrystallizations from a concentrated dichloromethane solution layered with hexane. The purity of these freshly prepared single crystals was confirmed by an elemental analysis.

A single-crystal X-ray diffraction (SCXRD) analysis showed **1–5** crystallizing as an isostructural series in the monoclinic space group $P2_1/n$ (No.14), with an asymmetric unit containing half of the dinuclear complexes and one CH_2Cl_2 molecule as the lattice solvent (Figure 1 and Table S1). Consequently, the unit cell comprises two complete dinuclear complexes and two lattice CH_2Cl_2 solvents. An inversion center is located at the midpoint of the central bis-bidentate Cl_2An^{2-} bridging ligand, rendering the two Ln^{III} centers equivalent by symmetry. In all five complexes, the coordination environments around the Ln^{III} centers are eight-coordinated with six N atoms from the two Tp^- capping ligands and two O atoms from the bridging Cl_2An^{2-} ligand. The average Ln–O and Ln–N distances of complexes **1–5** are in the ranges of 2.330(2) to 2.398(3) Å and 2.446(2) to 2.514(3) Å, respectively (Table 1), and are in agreement with previously reported values for other Ln-based complexes [10–18]. For all five complexes, the Ln–O distances are ≤0.12 Å shorter than the Ln-N distances, since the O atoms in the bridging Cl_2An^{2-} ligand exhibit a larger negative partial charge than the N atoms in the Tp^- ligand. A comparison of the bond distances in **1–5** reveals a slight decrease in the average Ln–O and Ln–N distances, from left to right across the isostructural series, as expected from the change in the ionic radii. This systematic decrease is evidence of the lanthanide contraction phenomenon [25–28].

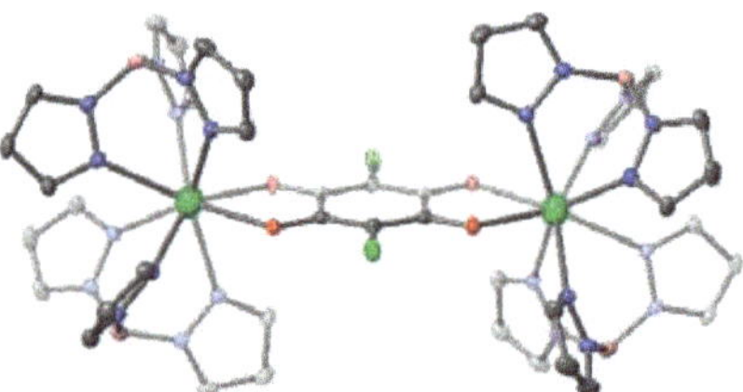

Figure 1. Solid state molecular structure of $[\{Ln(Tp)_2\}_2(\mu\text{-}Cl_2An)]\cdot2CH_2Cl_2$ with thermal ellipsoids drawn at a 50% probability level with the exception of lanthanide atoms shown at a 99% probability level. The spring green, blue, red, sundown, vivid lime green, and gray ellipsoids represent Ln, N, O, B, Cl, and C atoms, respectively. Hydrogen atoms and lattice solvent molecules are omitted for clarity.

Table 1. Selected bond distances [1] (Å) for **1–5**.

	Ln–O [2]	Ln–N [2]	Intramolecular Ln⋯Ln [3]	Intermolecular Ln⋯Ln [4]
1	2.398(3)	2.514(3)	8.7042(5)	8.7420(5)
2	2.375(3)	2.492(3)	8.661(2)	8.703(2)
3	2.363(2)	2.482(2)	8.6424(5)	8.7308(4)
4	2.348(2)	2.465(2)	8.6084(9)	8.699(1)
5	2.330(2)	2.446(2)	8.5599(9)	8.671(1)

[1] Note that the single crystal X-ray diffraction data for **1–5** were collected at different temperatures. [2] Averages of crystallographically independent Ln–O and six Ln–N values. [3] Symmetry code: $-x + 1$, $-y + 1$, $-z + 2$. [4] Closest separation.

The coordination geometry and environment around the Ln^{III} centers have a significant influence on the electronic structure and magnetic anisotropy. To determine the coordination geometries of the Ln^{III} centers for **1–5**, continuous shape measurements (CShM) were determined using the SHAPE Version 2.1 software [29,30], where two Ln^{III} centers are related by inversion symmetry and therefore possess the same coordination geometry (vide supra). Based on the resultant SHAPE indices (0.722–0.831 for **1–5**; Table 2), the eight-coordinate Ln^{III} centers of **1–5** are best described as having slightly distorted triangular dodecahedral geometries.

Table 2. Summary of SHAPE parameters [1] for lanthanide centers in the series of the dinuclear complexes **1–5**.

	SAPR [2]	TDD [3]	BTBR [4]
1	2.320	0.831	1.788
2	2.291	0.809	1.774
3	2.199	0.764	1.728
4	2.215	0.749	1.732
5	2.189	0.722	1.707

[1] A shape index equal to zero represents an ideal geometry. [2–4] SAPR, TDD, and BTBR are square antiprismatic, triangular dodecahedral, and bi-augmented trigonal prismatic geometries, respectively.

The bond distances within the quinoidal rings (e.g., Cl_2An^{2-} ligand) bound to the Ln^{III} centers provide strong information on the electronic structures of the ligands. The average C–O and C–C distances, with the delocalized bonding in the Cl_2An^{2-} rings of **1–5**, are in the range 1.252(4)–1.260(5) Å and 1.390(5)–1.394(4) Å, respectively, while the C–C distances with single bonding are in the range of 1.532(4)–1.542(4) (Table 3). The bond distances within the quinoidal rings of the bridging Cl_2An^{2-} ligands fall within the same error over **1–5**, which all adopt bi-separated delocalized forms [23]. These results are strongly supported by the infrared (IR) spectral data (Scheme 1).

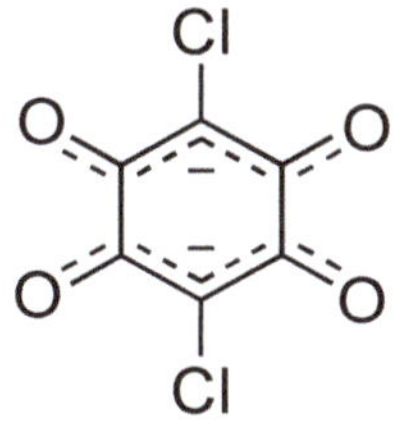

Scheme 1. Bi-separated delocalized structure of Cl_2An^{2-} in **1–5**.

Table 3. Selected bond distances [1] (Å) for the Cl_2An^{2-} moiety in **1–5**.

	C–O [2]	C–C [2]	C–C [3]
1	1.259(4)	1.391(5)	1.537(4)
2	1.252(4)	1.390(5)	1.542(4)
3	1.258(3)	1.394(3)	1.537(3)
4	1.256(3)	1.392(4)	1.534(4)
5	1.253(3)	1.393(4)	1.532(4)

[1] Bonds are shown in detail in Scheme 1. [2] Averages of two crystallographically independent delocalized C–O and C–C values. [3] Values of the C–C single bond moiety.

For complexes **1–5**, the respective intramolecular $Ln^{III}\cdots Ln^{III}$ separations through the Cl_2An^{2-} bridges are in the range of 8.5599(9)–8.7042(5) Å, whereas the closest intermolecular $Ln^{III}\cdots Ln^{III}$ separations, which are in the range of 8.671(1)–8.7420(5) Å, are comparable with the intramolecular distances (Table 1). These close intra- and intermolecular $Ln^{III}\cdots Ln^{III}$ distances may lead to magnetically dipolar interactions, which could create a small bias that allows for the quantum tunneling of the magnetization at the zero field (vide infra).

2.2. Infrared Spectroscopy

The IR spectra of **1–5** provide complementary structural feature information to that obtained by SCXRD analysis (Figure S1). The IR vibrations associated with the bridging Cl_2An^{2-} ligand in the five complexes are typified by predominant C–Cl and C–O vibrations at ~850 and 1530 cm^{-1}, respectively [22–24]. These results are a promising indication that the Cl_2An^{2-} ligand in complexes **1–5** is in a bi-separated delocalized form [22–24,31]. Furthermore, the presence of the ancillary Tp^- ligands in complexes **1–5** is confirmed by characteristic vibrations at ~2470 cm^{-1} (c.f., Tp^-, ν_{BH} = ~2440 cm^{-1}) [22–24,32]. The predominant contributions to the vibrational modes are from the skeleton of the ligand and, thus, there are no significant differences in the IR spectra. The changes in the atomic weight of the Ln^{III} centers are reflected in the specific peak shifts of the IR spectra. For the five complexes, the most notable change is observed in the Ln–O vibrations at ~460 cm^{-1}, where the Ln–O vibrational peak shifts to a higher energy (453–466 cm^{-1}) with an increasing atomic number. This increase can be explained by the lanthanide contraction effect.

2.3. Magnetic Properties

2.3.1. Static Magnetic Properties

For all of the dinuclear complex series (**1–5**), the temperature (T) dependence of the dc magnetic susceptibility (χ_M) data was collected under an applied dc field of 0.1 T in the temperature range of 1.8–300 K. A comparison of the resulting $\chi_M T$ versus T data reveals marked differences between the dc magnetism of complexes **1–5**. The $\chi_M T$ values at 300 K for **1–5** (15.73, 22.80, 27.97, 22.44, and 5.12 cm^3 K mol^{-1}, respectively) are in good agreement with the expected values (15.75, 23.63, 28.13, 22.95, and 5.14 cm^3 K mol^{-1}, respectively) for two non-interacting Ln^{III} centers. The $\chi_M T$ products for complexes **2–5** gradually decreased over the temperature range of 300–50 K. Subsequently, they rapidly decreased below 50 K and finally reached values of 15.02, 8.91, 4.47, 13.32, and 2.50 cm^3 K mol^{-1}, respectively, at 1.8 K, due to the depopulation of the excited crystal field state. For complexes **2–5**, the isothermal dc magnetization at 1.8 K increased steeply with an increasing magnetic field at low magnetic field regions before increasing linearly in high magnetic field regions, finally reaching respective values of 9.63, 11.51, 9.32, and 3.85 μ_B at 7 T without saturation, indicating very strong magnetic anisotropy (Figure 2b). Moreover, no hysteresis was observed for **2–5**, even at 1.8 K using a conventional superconducting quantum interference device (SQUID). This suggests that the static magnetic behavior observed for **2–5** arose from significant spin-orbit coupling interactions and a strong unquenched orbital angular momentum. On the other hand, **1** comprises a half-filled 4f shell. Thus, its ground state has no orbital angular momentum and can be considered a spin-only system. In this case, the first-order effect of the spin-orbit coupling disappears for the ground electronic term $^8S_{7/2}$, and magnetic anisotropy is caused by the second-order effect of the spin-orbit coupling; this is known as zero-field splitting. The $\chi_M T$ value of 15.73 cm^3 K mol^{-1} at 300 K observed for **1** is in good agreement with the expected values (15.75 cm^3 K mol^{-1}) for two non-interacting Gd^{III} centers. The $\chi_M T$ product for **1** remains essentially constant down to 10 K and then gradually decreases to 15.02 cm^3 K mol^{-1} at 1.8 K. The magnetization curve of **1** at 1.8 K was saturated at 14 μ_B at 7 T, confirming that **1** is in the ground state of $^8S_{7/2}$.

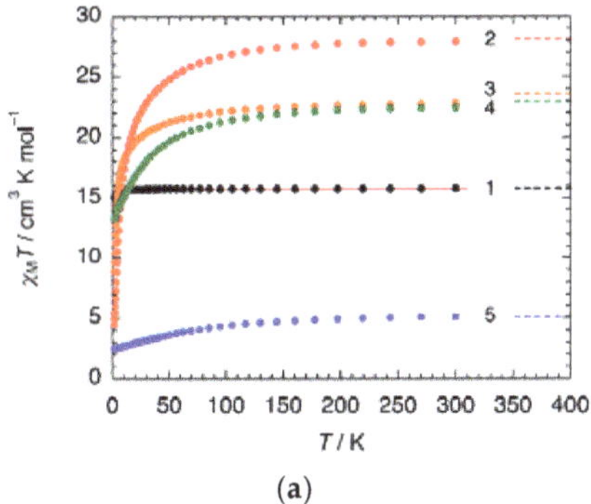
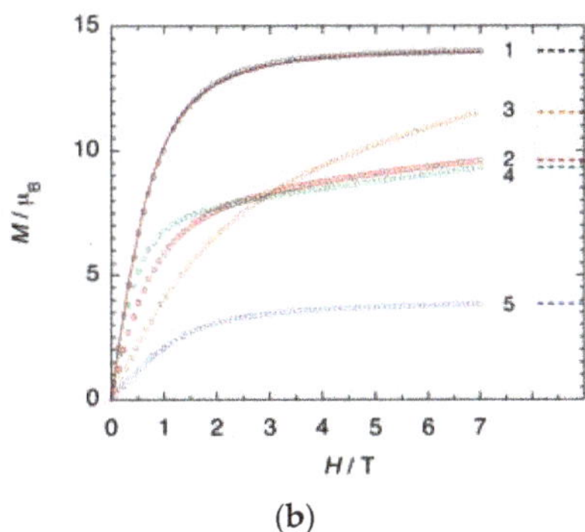

(a)   (b)

Figure 2. (a) Temperature dependence of the molar magnetic susceptibility times the temperature for **1–5** over the temperature range between 1.8 and 300 K under an applied dc field of 0.1 T; (b) Field dependence of molar magnetization for **1–5** over the dc field range between 0 and 7 T at 1.8 K. The broken lines correspond to ideal free-ion values. The solid red lines represent fits to the experimental data using the spin Hamiltonian based on the zero-field splitting, which has been previously described in detail [23].

2.3.2. Dynamic Magnetic Properties

To study the possibility of a slow magnetic relaxation, the ac susceptibility measurements for **1–5** were performed at 1.8 K with a dc magnetic field in the range of 0–0.3 T. The out-of-phase ac susceptibility (χ_M'') signals for all five complexes in the absence of an applied field did not present any apparent peaks in the available frequency (ν) range. As expected for the eight-coordinated triangular dodecahedral geometry of the two Ln^{III} centers, an approximate D_{2d} symmetry was observed. This led to the crystal field parameters B^0_2, B^0_4, B^4_4, B^0_6, and B^4_6, wherein B^4_4 and B^4_6 are the off-diagonal components. The existence of these off-diagonal crystal field parameters strongly suggests the mixing of the ground M_J states. Furthermore, the Ln^{III} centers in **1–5** comprise isotopes that display a nuclear spin, resulting in the nuclear hyperfine interaction effect [33–37]. Additionally, the intra- and/or intermolecular separations between the Ln^{III} centers suggests the presence of dipolar interactions [10–18]. These contributions lead to the absence of a slow magnetic relaxation under a zero applied dc field, thereby allowing the quantum tunneling of the magnetization. In such cases, the application of a dc field can suppress and break up the quantum tunneling, caused by nuclear hyperfine couplings, dipolar interactions, and transverse fields from the off-diagonal crystal field splittings, and reveal the slow relaxation of the magnetization. However, under small static dc magnetic fields, frequency-dependent non-zero χ_M'' signals were only clearly observed for the Kramers ions in complexes **4** and **5** (Figure 3, Figures S2 and S3). For these two complexes, each χ_M'' peak maximum shifted to a lower frequency with an increasing applied dc field ≤0.1 T. A further increase in the applied dc field resulted in the maximum χ_M'' shifting to a higher frequency. Notably, **4** and **5** also presents another minor magnetic relaxation process at higher dc fields (Figure 3), which may arise from thermally assisted quantum tunneling [10–18]. The dc field dependence of the low temperature relaxation times ($\tau = 1/2\pi\nu$) for **4** and **5** was extracted at each of these fields by fitting ν versus χ_M' and χ_M'' and the Argand plots [38,39] (χ_M' versus χ_M'') using a generalized Debye model [40]. The values determined for **4** and **5** are listed in Tables S2 and S3 and plotted in Figure S3. The two magnetic relaxation processes of the direct and quantum relaxation pathways were elucidated by fitting the variable-field magnetic relaxation data of **4** and **5** using Equation (1) [10–18]:

$$\tau^{-1} = AH^nT + \frac{B_1}{1 + B_2H^2} \tag{1}$$

where the first and second terms represent the respective direct and quantum tunneling pathways. Moreover, because of the presence of Kramers ions, the power index m = 4 was used for the direct process. The best fits, based on Equation (1), are presented as solid black lines in Figure 3 and are summarized in Table 4. These results imply that the dipolar interactions and/or off-diagonal crystal fields support quantum tunneling at low magnetic dc fields. On the other hand, due to the presence of spin-active nuclei, single-phonon direct relaxation dominates at high dc fields. In addition, the optimum dc field for **4** and **5** was determined as ~0.1 T.

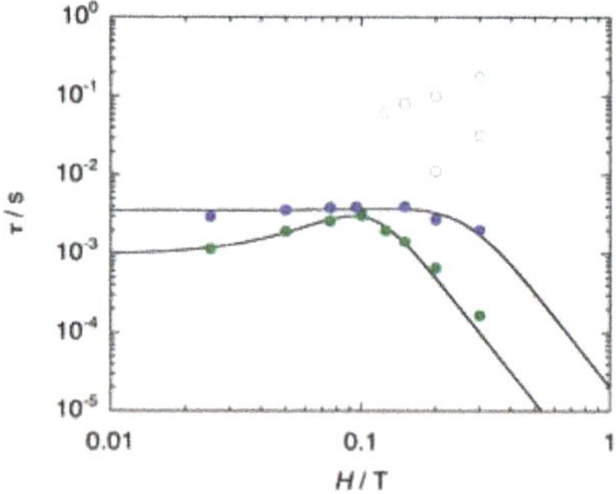

Figure 3. Comparison of the field-dependent relaxation time (τ) at 1.8 K for two distinct major (filled circles) and minor (open circles) relaxations observed for **4** (green), and **5** (blue). The solid black lines represent fits to the experimental data using the appropriate magnetic relaxation pathways considering both the quantum tunneling and direct processes (Equation (1)).

Table 4. Summary of field-dependent ac magnetic data [1] for 4 and 5.

	A (s^{-1} K^{-1} T^{-4})	B_1 (s^{-1})	B_2 (T^{-2})
4	6.80×10^5	1.03×10^3	3.85×10^2
5	2.70×10^4	2.82×10^2	6.42

[1] Data measured at 1.8 K.

Subsequently, ac susceptibility measurements were performed under an applied dc field of 0.1 T in the temperature range of 1.8–10 K (Figure 4), where the optimum dc magnetic field for **4** and **5** was determined as 0.1 T (variable-field magnetic relaxation data; vide supra, Figure 3). The temperature dependences of the magnetic relaxation times for **4** and **5** were extracted in the temperature range of 1.8–5.0 K by fitting ν versus χ_M' and χ_M'' and the Argand plots using a generalized Debye model (Figures 4 and 5, and Tables S4 and S5, respectively). The Argand plots for **4** and **5** comprised one semicircle with small α parameters in the ranges of 0.08–0.42 (**4**) and of 0.02–0.18 (**5**).

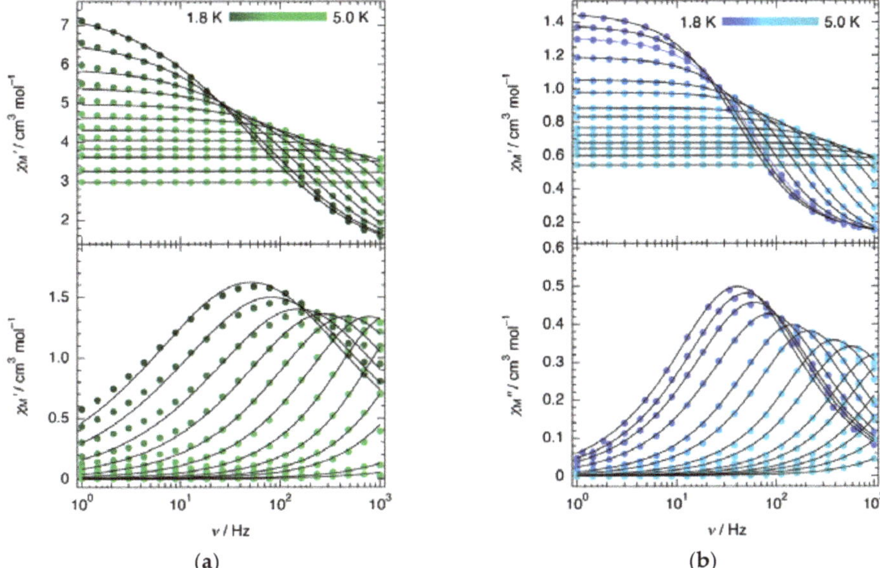

Figure 4. Frequency dependence of the molar in-phase (top) and out-of-phase (bottom) susceptibility for (**a**) **4** and (**b**) **5** over the frequency 1–1000 Hz and the temperature range 1.8–5.0 K in a 2.5 Oe ac field under an applied dc field of 0.1 T. The solid black lines represent fits to the experimental data using the generalized Debye model with the α parameter in the ranges of 0.08–0.42 for **4** and of 0.02–0.18 for **5**.

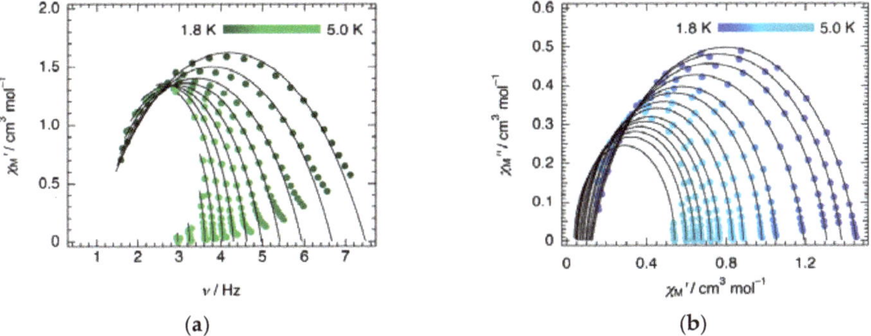

Figure 5. Argand plots for the molar ac susceptibility data of (**a**) **4** and (**b**) **5** in the temperature range 1.8–5.0 K under an applied dc field of 0.1 T. The solid black lines correspond to fits to the experimental data using the generalized Debye model with the α parameter in the ranges of 0.08–0.42 for **4** and of 0.02–0.18 for **5**.

The linear regions at a high temperature in the plots of τ versus $1/T$ (Figure S4) were fitted, assuming an Orbach relaxation (ideal thermal excitation over the energy barrier for the molecule [10–18,40]), as described by Equation (2):

$$\tau^{-1} = \tau_0^{-1} \exp\left(-\frac{\Delta_{\text{eff}}}{k_{\text{B}}T}\right) \tag{2}$$

Arrhenius fits of the temperature-dependent relaxation time afford the thermally activated barriers $\Delta_{\text{eff}} = 26.0\ \text{cm}^{-1}$ ($\tau_0 = 1.79 \times 10^{-9}$ s) for **4** and $21.5\ \text{cm}^{-1}$ for **5** ($\tau_0 = 2.81 \times 10^{-8}$ s). The extracted τ_0 values fall within the typical range of Ln^{III}-based single-molecule and single-ion magnets [10–18]. As the temperature decreases, the plots of τ versus $1/T$ for **4** and **5** become gradually nonlinear (Figure 6). Such behavior suggests the coexistence of multiple magnetic relaxation pathways, which is caused by energy transfer from the spin to the lattice; this is known as the spin-lattice relaxation [10–18]. Hence, the variable temperature relaxation times for **4** and **5** were analyzed in terms of their spin-lattice relaxation (Equation (3)):

$$\tau^{-1} = AH^nT + \frac{B_1}{1 + B_2H^2} + \tau_0^{-1}\exp\left(-\frac{\Delta_{\mathrm{eff}}}{k_BT}\right) + CT^m \tag{3}$$

where the first, second, third, and fourth terms represent the direct, quantum tunneling, Orbach, and Raman relaxation processes, respectively [10–18]. Since the temperature dependence of the τ data was collected at the optimum dc field of 0.1 T, the direct and quantum tunneling contributions should be excluded. Therefore, the overall τ versus $1/T$ data for **4** and **5** can only be fit with the Orbach and Raman contributions. The best fits, presented as solid black lines, are illustrated in Figure 6 and Figure S5, while their best-fit parameters are listed in Table 5. The calculated m values are smaller than the ideal value of $m = 9$ for the Kramers ions, suggesting that these Raman-like relaxations are attributed to acoustic and optical vibrations [10–18].

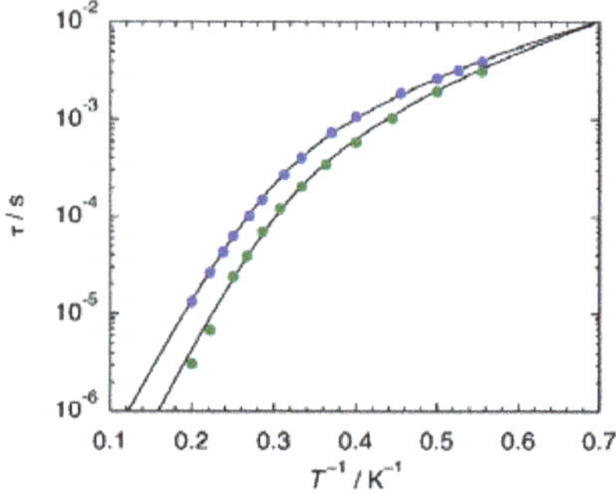

Figure 6. Comparison of the temperature-dependent relaxation under an applied dc field of 0.1 T for **4** (green) and **5** (blue). The solid black lines represent fits to the experimental data using the appropriate magnetic relaxation pathways (Equation (3)).

Table 5. Summary of temperature-dependent ac magnetic data for **4** and **5**.

	τ_0 (s)	Δ_{eff} (cm^{-1})	C (s^{-1} K^{-n})	m
4	3.04×10^{-9}	25.9	17.50	4.84
5	2.68×10^{-8}	22.3	31.39	3.55

2.4. Electrochemistry

The Cl_2An^{2-} ligand could be utilized not only as a bridging unit for designing novel multinuclear coordination assemblies, but also as a non-innocent ligand, namely a reversible redox active ligand. Therefore, utilizing the non-innocent Cl_2An^{2-} as the bridging ligand in SMMs offers the enticing possibility for redox controllable magnetic behavior via an electrical signal. To probe the redox behavior of the field-induced SMMs **4** and **5**, electrochemical measurements were carried out in degassed CH_2Cl_2 with n-Bu$_4$NPF$_6$ as the supporting electrolytes. The cyclic voltammogram shows the single quasi-reversible one-electron reduction at $E_{1/2} = 1.05$ V for **4** and 1.07 V for **5** versus the ferrocene/ferrocenium couple (vs. Fe(Cp)$_2$$^{0/1+}$), which was assigned as the ligand-based process (Figure S6). The $E_{1/2}$ values of **4** and **5** are similar to those reported for related compounds [22–24]. Attempts and efforts to isolate the one-electron reduced products can be found elsewhere [22–24].

3. Experimental Section

3.1. Materials and Methods

The lanthanide chlorides and solvents were purchased from Wako Pure Chemical Industries, Ltd. (Osaka, Japan). $Na_2Cl_2An\cdot3H_2O$ and KTp were purchased from Tokyo Chemical Industry (TCI) Co., Ltd. (Tokyo, Japan). All chemicals were of reagent grade and were used as received. Both CH_2Cl_2 and hexane were of super dehydrated grade. All of the reactions and manipulations were performed under aerobic conditions at an ambient temperature.

3.2. Synthesis of [{Ln(Tp)$_2$}$_2$(μ-Cl$_2$An)]·2CH$_2$Cl$_2$ (Ln = Gd (1), Tb (2), Ho (3), Er (4) and Yb (5))

All the complexes were synthesized according to a previously reported method [21], with modifications. Single crystals suitable for single-crystal X-ray measurements were obtained by recrystallization from CH_2Cl_2/hexane. Notably, several recrystallizations were necessary to obtain samples of sufficient purity. Crystalline yields for each complex were in the range of 38–51%. Anal. Calcd. for $C_{44}H_{44}B_4Cl_6Gd_2N_{24}O_4$: C, 34.24; H, 2.87; N, 21.78%. Found: C, 34.56; H, 2.91, N, 21.94%. Anal. Calcd. for $C_{44}H_{44}B_4Cl_6Tb_2N_{24}O_4$: C, 34.17; H, 2.87; N, 21.73%. Found: C, 34.29; H, 2.88, N, 21.63%. Anal. Calcd. for $C_{44}H_{44}B_4Cl_6Ho_2N_{24}O_4$: C, 33.90; H, 2.85; N, 21.57%. Found: C, 34.12; H, 2.71, N, 21.66%. Anal. Calcd. for $C_{44}H_{44}B_4Cl_6Er_2N_{24}O_4$: C, 33.80; H, 2.84; N, 21.50%. Found: C, 33.81; H, 2.98, N, 21.24%. Anal. Calcd. for $C_{44}H_{44}B_4Cl_6Yb_2N_{24}O_4$: C, 33.55; H, 2.82; N, 21.34%. Found: C, 33.42; H, 2.91, N, 21.55%.

3.3. Single Crystal X-ray Crystallography

The single crystals of **2–5** were coated with Nujol, quickly mounted on MicroLoops (MiTeGen LLC., Ithaca, NY, USA), and immediately cooled in a cold N_2 stream to prevent any lattice solvent loss. The data collections were performed on a Rigaku Saturn 724 or R-AXIS RAPID II IP diffractometer (Rigaku Corporation, Tokyo Japan) with graphite-monochromated Mo-$K\alpha$ radiation ($\lambda = 0.71075$ Å) and a low-temperature device. The data integration, preliminary data analysis, and absorption collections were performed on a Rigaku CrystalClear-SM 1.4.0 SP1 [41], using the CrystalStructure 4.2.2 [42] crystallographic software packages. The molecular structures were solved by the direct methods included in SIR2011 [43] and refined with the SHELXL [44] program. All non-hydrogen atoms were refined anisotropically. CCDC-1905608–1905611 for **2–5** contain the supplementary crystallographic data for this paper and can be obtained free of charge from The Cambridge Crystallographic Data Centre via www.ccdc.cam.ac.uk/data_request/cif. All the hydrogen atoms were included in the calculated positions. Table S1 summarizes the lattice constants and structure refinement parameters for complexes **2–5**.

3.4. Physical Measurements

The elemental analysis was performed on a J-Science Lab Micro Corder JM10 (J-Science Lab Co., Ltd., Kyoto, Japan). The Fourier transform infrared spectra were collected using KBr disks, on a JASCO FT/IR-410 spectrometer (JASCO Corporation, Tokyo, Japan) in the range of 400–4000 cm^{-1} at a resolution of 4 cm^{-1} at an ambient temperature. The magnetic data were collected using a Quantum Design MPMS3 SQUID magnetometer (Quantum Design Japan, Inc., Tokyo Japan). The measurements were performed with crushed crystalline samples in a calibrated gelatin capsule. The dc magnetic susceptibility measurements were performed in the temperature range of 1.8–300 K in a dc field of 0.1 T. The field-dependent dc magnetization measurements were performed from −7 to +7 T at 1.8 K. The ac susceptibility measurements were performed in the temperature range of 1.8–15 K in a 2.5 Oe ac field, oscillating at a frequency range of 1–997 Hz in different applied dc fields. The obtained magnetic susceptibility data were corrected for diamagnetic contributions from the sample holder as well as for the core diamagnetism of each sample, estimated from Pascal's constants [45]. The cyclic voltammetric measurements were performed in a 0.1 M CH_2Cl_2 solution of n-Bu$_4$NPF$_6$ using an

ALS/chi Electrochemical Analyzer Model 610A with a computer-controlled workstation (ALS Co., Ltd, Tokyo, Japan). The solutions contained approximately 1 mM in compounds. The experiments were performed under a continuous flow of N_2 gas using a standard three-electrode cell (platinum working and counter electrodes with an Ag/Ag^+ reference electrode, respectively). The reported potentials are all referenced to the $Fe(Cp)_2^{0/1+}$ couple, which was determined using $Fe(Cp)$ as an internal standard at 0 V.

4. Conclusions and Outlook

The series of Cl_2An^{2-} bridged dinuclear Ln complexes with the formula $[\{Ln(Tp)_2\}_2(\mu\text{-}Cl_2An)]\cdot2CH_2Cl_2$ (Ln = Gd (**1**), Tb (**2**), Ho (**3**), Er (**4**), and Yb (**5**)) were successfully synthesized and systematically characterized by a single X-ray diffraction and by SQUID measurements. All five dinuclear Ln complexes were isostructural and clearly displayed the structural change attributed to the lanthanide contraction effect. A comparison of the dc magnetic data for **1–5** revealed clear differences depending on the Ln^{III} centers. None of the five complexes displayed any slow relaxation of the magnetization under a zero applied dc field, while only two complexes (**4** and **5**) presented a slow relaxation of the magnetization in the presence of small dc fields. These two complexes correspond to Kramers ions. The dynamic magnetic properties of **4** and **5** were interpreted by using multiple relaxation pathways, whereby both Orbach and Raman relaxation processes were considered.

The proposed series comprises dinuclear Ln complexes with an electroactive Cl_2An^{2-} bridging ligand. For more potential applications, an important challenge is to switch the slow magnetization phenomena with chemical and physical external fields. These electrochemical molecular switches must be from particularly attractive molecule-based devices, in which the electroactive molecules are reversibly converted between different redox states triggered by an electrical signal. To realize such applications, electrical switchable characteristics focusing on complexes **4** and **5** are in progress.

Supplementary Materials: The following are available online at http://www.mdpi.com/2312-7481/5/2/30/s1, Table S1: X-ray crystallographic data for **1–5**, Figure S1: IR spectra of **1–5**, Figure S2: Frequency dependence of ac susceptibilities under variable dc fields for **1–5**, Figure S3: Argand plots under variable dc fields for **1–5**, Figure S4: τ versus $1/T$ plots with fits using Equation (2) for **4** and **5**, Figure S5: τ versus $1/T$ plots with fits using Equation (3) for **4** and **5**, Table S2: Summary of dc magnetic fields dependent relaxation times and α values for **4**, Table S3: Summary of dc magnetic fields dependent relaxation times and α values for **5**, Table S4: Summary of temperature dependent relaxation times and α values for **4**, Table S5: Summary of temperature dependent relaxation times and α values for **5**, Figure S6: Cyclic voltammograms of **4** and **5**.

Author Contributions: R.I. conceived and designed the experiment. R.I. wrote the manuscript in consultation with co-authors; R.I., T.N., S.M., and S.K. executed syntheses and their characterization, and single-crystal X-ray diffraction measurements and their structure refinement; R.I., K.K., and M.Y. executed magnetic measurements. Ryuta Ishikawa analyzed all magnetic data.

Funding: This work was supported financially by the Ministry of Education, Culture, Sports, Science and Technology (MEXT) KAKENHI (Grant-in-Aid for Scientific Research on Innovative Areas), Grant Number 18H04529 "Soft Crystals" (Ryuta Ishikawa) and 17H05390 "Coordination Asymmetry" (Satoshi Kawata), as well as by the Japan Society for the Promotion of Science (JSPS) KAKENHI, Grant Number (Grant-in-Aid for Scientific Research (C)) 16K05735 (Satoshi Kawata). This work was also supported financially by the Central Research Institute of Fukuoka University, Grant Number 171041 (Ryuta Ishikawa) and 171011 (Satoshi Kawata).

Acknowledgments: Masahiro Yamashita thanks the support by the 111 project (B18030) from China. We would like to thank Editage (www.editage.jp) for English language editing.

Conflicts of Interest: The authors declare no conflict of interest.

References

1. Mannini, M.; Pineider, F.; Sainctavit, P.; Danieli, C.; Otero, E.; Sciancalepore, C.; Talarico, A.M.; Arrio, M.-A.; Cornia, A.; Gatteschi, D.; et al. Magnetic memory of a single-molecule quantum magnet wired to a gold surface. *Nat. Mater.* **2009**, *8*, 194–197. [CrossRef]
2. Leuenberger, M.N.; Loss, D. Quantum computing in molecular magnets. *Nature* **2001**, *410*, 789–793. [CrossRef]

3. Meier, F.; Levy, J.; Loss, D. Quantum computing with spin cluster qubits. *Phys. Rev. Lett.* **2003**, *90*, 47901–47904. [CrossRef] [PubMed]
4. Winpenny, R.E.P. Quantum information processing using molecular nanomagnets as qubits. *Angew. Chem. Int. Ed.* **2008**, *47*, 7992–7994. [CrossRef]
5. Chiesa, A.; Whitehead, G.F.S.; Carretta, S.; Carthy, L.; Timco, G.A.; Teat, S.J.; Amoretti, G.; Pavarini, E.; Winpenny, R.E.P.; Santini, P. Molecular nanomagnets with switchable coupling for quantum simulation. *Sci. Rep.* **2014**, *4*, 7423. [CrossRef] [PubMed]
6. Bogani, L.; Wernsdorfer, W. Molecular spintronics using single-molecule magnets. *Nat. Mater.* **2008**, *7*, 179–186. [CrossRef]
7. Camarero, J.; Coronado, E. Molecular vs. inorganic spintronics: The role of molecular materials and single molecules. *J. Mater. Chem.* **2009**, *19*, 1678–1684. [CrossRef]
8. Sanvito, S. Molecular spintronics. *Chem. Soc. Rev.* **2011**, *40*, 3336–3355. [CrossRef]
9. Clemente-Juan, J.M.; Coronado, E.; Gaita-Ariño, A. Magnetic polyoxometalates: From molecular magnetism to molecular spintronics and quantum computing. *Chem. Soc. Rev.* **2012**, *41*, 7464–7478. [CrossRef]
10. Rinehart, J.D.; Long, J.R. Exploiting single-ion anisotropy in the design of f-element single-molecule magnets. *Chem. Sci.* **2011**, *2*, 2078–2085. [CrossRef]
11. Sorace, L.; Benelli, C.; Gatteschi, D. Lanthanides in molecular magnetism: Old tools in a new field. *Chem. Soc. Rev.* **2011**, *40*, 3092–3104. [CrossRef]
12. Woodruff, D.N.; Winpenny, R.E.P.; Layfield, R.A. Lanthanide single-molecule magnets. *Chem. Rev.* **2013**, *113*, 5110–5148. [CrossRef] [PubMed]
13. Habib, F.; Murugesu, M. Lessons learned from dinuclear lanthanide nano-magnets. *Chem. Soc. Rev.* **2013**, *42*, 3278–3288. [CrossRef]
14. Liddle, S.T.; Van Slageren, J. Improving f-element single molecule magnets. *Chem. Soc. Rev.* **2015**, *44*, 6655–6669. [CrossRef] [PubMed]
15. Pointillart, F.; Cador, O.; Le Guennic, B.; Ouahab, L. Uncommon lanthanide ions in purely 4f Single Molecule Magnets. *Coord. Chem. Rev.* **2017**, *346*, 150–175. [CrossRef]
16. McAdams, S.G.; Ariciu, A.-M.; Kostopoulos, A.K.; Walsh, J.P.S.; Tuna, F. Molecular single-ion magnets based on lanthanides and actinides: Design considerations and new advances in the context of quantum technologies. *Coord. Chem. Rev.* **2017**, *346*, 216–239. [CrossRef]
17. Dey, A.; Kalita, P.; Chandrasekhar, V. Lanthanide(III)-based single-ion magnets. *ACS Omega* **2018**, *3*, 9462–9475. [CrossRef]
18. Zhu, Z.; Guo, M.; Li, X.-L.; Tang, J. Molecular magnetism of lanthanide: Advances and perspectives. *Coord. Chem. Rev.* **2019**, *378*, 350–364. [CrossRef]
19. Sessoli, R.; Tsai, H.L.; Schake, A.R.; Wang, S.; Vincent, J.B.; Folting, K.; Gatteschi, D.; Christou, G.; Hendrickson, D.N. High-spin molecules: [Mn$_{12}$O$_{12}$(O$_2$CR)$_{16}$(H$_2$O)$_4$]. *J. Am. Chem. Soc.* **1993**, *115*, 1804–1816. [CrossRef]
20. Sessoli, R.; Gatteschi, D.; Caneschi, A.; Novak, M.A. Magnetic bistability in a metal-ion cluster. *Nature* **1993**, *365*, 141–143. [CrossRef]
21. Abdus Subhan, M.; Kawahata, R.; Nakata, H.; Fuyuhiro, A.; Tsukuda, T.; Kaizaki, S. Synthesis, structure and spectroscopic properties of chloranilate-bridged 4f-4f dinuclear complexes: A comparative study of the emission properties with Cr-Ln complexes. *Inorg. Chim. Acta* **2004**, *357*, 3139–3146. [CrossRef]
22. Dunstan, M.A.; Rousset, E.; Boulon, M.-E.; Gable, R.W.; Sorace, L.; Boskovic, C. Slow magnetisation relaxation in tetraoxolene-bridged rare earth complexes. *Dalton Trans* **2017**, *46*, 13756–13767. [CrossRef] [PubMed]
23. Ishikawa, R.; Michiwaki, S.; Noda, T.; Katoh, K.; Yamashita, M.; Matsubara, K.; Kawata, S. Field-induced slow magnetic relaxation of mono- and dinuclear dysprosium(III) complexes coordinated by a chloranilate with different resonance forms. *Inorganics* **2018**, *6*, 7. [CrossRef]
24. Zhang, P.; Perfetti, M.; Kern, M.; Hallmen, P.P.; Ungur, L.; Lenz, S.; Ringenberg, M.R.; Frey, W.; Stoll, H.; Rauhut, G.; et al. Exchange coupling and single molecule magnetism in redox-active tetraoxolene-bridged dilanthanide complexes. *Chem. Sci.* **2018**, *9*, 1221–1230. [CrossRef] [PubMed]
25. Flanagan, B.M.; Bernhardt, P.V.; Krausz, E.R.; Lüthi, S.R.; Riley, M.J. A ligand-field analysis of the trensal (H$_3$trensal = 2,2′,2″-Tris(salicylideneimino)triethylamine) ligand. An application of the angular overlap model to lanthanides. *Inorg. Chem.* **2002**, *41*, 5024–5033. [CrossRef]

26. Quadrelli, E.A. Lanthanide contraction over the 4f series follows a quadratic decay. *Inorg. Chem.* **2002**, *41*, 167–169. [CrossRef] [PubMed]

27. Seitz, M.; Oliver, A.G.; Raymond, K.N. The lanthanide contraction revisited. *J. Am. Chem. Soc.* **2007**, *129*, 11153–11160. [CrossRef]

28. D'Angelo, P.; Zitolo, A.; Migliorati, V.; Chillemi, G.; Duvail, M.; Vitorge, P.; Abadie, S.; Spezia, R. Revised ionic radii of lanthanoid(III) ions in aqueous solution. *Inorg. Chem.* **2011**, *50*, 4572–4579. [CrossRef] [PubMed]

29. Casanova, D.; Llunell, M.; Alemany, P.; Alvarez, S. The rich stereochemistry of eight-vertex polyhedra: A continuous shape measures study. *Chem. Eur. J.* **2005**, *11*, 1479–1494. [CrossRef] [PubMed]

30. Llunell, M.; Casanova, D.; Cirera, J.; Bofill, J.M.; Alemany, P.; Alvarez, S. *SHAPE*; Version 2.1; University of Barcelona: Barcelona, Spain, 2013.

31. Kitagawa, S.; Kawata, S. Coordination compounds of 1,4-dihydroxybenzoquinone and its homologues. Structures and properties. *Coord. Chem. Rev.* **2002**, *224*, 11–34. [CrossRef]

32. Apostolidis, C.; Rebizant, J.; Kanellakopulos, B.; von Ammon, R.; Dornberger, E.; Müller, J.; Powietzka, B.; Nuber, B. Homoscorpionates (hydridotris(1-pyrazolyl)borato complexes) of the trivalent 4f ions. The crystal and molecular structure of [(HB(N$_2$C$_3$H$_3$)$_3$)$_3$Ln^{III}, (Ln = Pr, Nd). *Polyhedron* **1997**, *16*, 1057–1068. [CrossRef]

33. Sears, V.F. Neutron scattering lengths and cross sections. *Neutron News* **1992**, *3*, 26–37. [CrossRef]

34. Chen, Y.-C.; Liu, J.-L.; Wernsdorfer, W.; Liu, D.; Chibotaru, L.F.; Chen, X.-M.; Tong, M.-L. Hyperfine-interaction-driven suppression of quantum tunneling at zero field in a holmium(III) single-ion magnet. *Angew. Chem. Int. Ed.* **2017**, *56*, 4996–5000. [CrossRef]

35. Moreno-Pineda, E.; Damjanović, M.; Fuhr, O.; Wernsdorfer, W.; Ruben, M. Nuclear spin isomers: Engineering a Et$_4$N[DyPc$_2$] spin qudit. *Angew. Chem. Int. Ed.* **2017**, *56*, 9915–9919. [CrossRef]

36. Kishi, Y.; Pointillart, F.; Lefeuvre, B.; Riobé, F.; Le Guennic, B.; Golhen, S.; Cador, O.; Maury, O.; Fujiwara, H.; Ouahab, L. Isotopically enriched polymorphs of dysprosium single molecule magnets. *Chem. Commun.* **2017**, *53*, 3575–3578. [CrossRef]

37. Flores Gonzalez, J.; Pointillart, F.; Cador, O. Hyperfine coupling and slow magnetic relaxation in isotopically enriched Dy^{III} mononuclear single-molecule magnets. *Inorg. Chem. Front.* **2019**. [CrossRef]

38. Cole, K.S.; Cole, R.H. Dispersion and absorption in dielectrics, I. alternating current characteristics. *J. Chem. Phys.* **1941**, *9*, 341–351. [CrossRef]

39. Guo, Y.-N.; Xu, G.-F.; Guo, Y.; Tang, J. Relaxation dynamics of dysprosium(iii) single molecule magnets. *Dalton Trans.* **2011**, *40*, 9953–9963. [CrossRef] [PubMed]

40. Gatteschi, D.; Sessoli, R.; Villain, J. *Molecular Nanomagnets*; Oxford University Press: Oxford, UK, 2006. [CrossRef]

41. *CrystalClear-SM*; Version 1.4.0 SP1; Rigaku and Rigaku/MSC: The Woodlands, TX, USA, 2008.

42. *CrystalStructure*; Version 4.2.2; Rigaku and Rigaku/MSC: The Woodlands, TX, USA, 2017.

43. Burla, M.C.; Caliandro, R.; Camalli, M.; Carrozzini, B.; Cascarano, G.L.; Giacovazzo, C.; Mallamo, M.; Mazzone, A.; Polidori, G.; Spagna, R. SIR2011: A new package for crystal structure determination and refinement. *J. Appl. Crystallogr.* **2012**, *45*, 357–361. [CrossRef]

44. Sheldrick, G.M. Crystal structure refinement with SHELXL. *Acta Crystallogr. Sect. C* **2015**, *71*, 3–8. [CrossRef]

45. Bain, G.A.; Berry, J.F. Diamagnetic corrections and Pascal's constants. *J. Chem. Educ.* **2008**, *85*, 532–536. [CrossRef]

 *magnetochemistry*

Article

How to Quench Ferromagnetic Ordering in a CN-Bridged Ni(II)-Nb(IV) Molecular Magnet? A Combined High-Pressure Single-Crystal X-Ray Diffraction and Magnetic Study

Gabriela Handzlik [1] , Barbara Sieklucka [1] , Hanna Tomkowiak [2] , Andrzej Katrusiak [2] and Dawid Pinkowicz [1,*]

[1] Faculty of Chemistry, Jagiellonian University, Gronostajowa 2, 30-387 Kraków, Poland; gabriela.handzlik@uj.edu.pl (G.H.); barbara.sieklucka@uj.edu.pl (B.S.)
[2] Faculty of Chemistry, Adam Mickiewicz University, Umultowska 89b, 61-614, Poznań, Poland; hannat@amu.edu.pl (H.T.); katran@amu.edu.pl (A.K.)
* Correspondence: dawid.pinkowicz@uj.edu.pl; Tel.: +48-12-686-2457

Received: 12 April 2019; Accepted: 28 May 2019; Published: 1 June 2019

Abstract: High-pressure (HP) structural and magnetic properties of a magnetic coordination polymer $\{[Ni^{II}(pyrazole)_4]_2[Nb^{IV}(CN)_8]\cdot4H_2O\}_n$ (Ni$_2$Nb) are presented, discussed and compared with its two previously reported analogs $\{[Mn^{II}(pyrazole)_4]_2[Nb^{IV}(CN)_8]\cdot4H_2O\}_n$ (Mn$_2$Nb) and $\{[Fe^{II}(pyrazole)_4]_2[Nb^{IV}(CN)_8]\cdot4H_2O\}_n$ (Fe$_2$Nb). Ni$_2$Nb shows a significant decrease of the long-range ferromagnetic ordering under high pressure when compared to Mn$_2$Nb, where the pressure enhances the T_c (magnetic ordering temperature), or to Fe$_2$Nb exhibiting a pressure-induced spin crossover. The different HP magnetic responses of the three compounds were rationalized and correlated with the structural models as determined by single-crystal X-ray diffraction.

Keywords: ferromagnetism; long-range magnetic ordering; X-ray diffraction; high pressure; nickel(II); octacyanidoniobate(IV); coordination polymers

1. Introduction

Structural and magnetic measurements under high pressure are the most reliable source of straightforward magneto-structural correlations in crystalline magnetic solids. These types of studies make it possible to fine-tune the structure and physical properties in a continuous manner—a feature that cannot be achieved via chemical modifications, which often introduce unexpected complications (different packing modes, additional intra- and intermolecular contacts). The application of high pressure is known to be extremely useful for enhancing the magnetic ordering temperature of extended coordination systems [1–5], ligand field and magnetic anisotropy tuning of mononuclear complexes [6,7] and control of the spin crossover behavior [8,9]. The combined high-pressure single-crystal X-ray diffraction (scXRD) and SQUID magnetometry make a perfect set of tools to study and understand the changes induced by this type of mechanical stimulus. scXRD structural analysis experiments are commonly performed using diamond anvil cells (DACs) [10] and, in the case of high-pressure SQUID magnetometry, the most common environment chamber is a piston-cylinder cell (PCC) made of diamagnetic copper-beryllium alloy [11].

Magnetic coordination polymers are known for their high responsiveness to mechanical stress and high pressure. In particular, cyanide-bridged Prussian Blue analogs [12] and the related octacyanometallate-based bimetallic assemblies [2] show significant magnetic changes under high pressure.

Herein, we present the quenching of the long-range ferromagnetic ordering in a cyanide-bridged coordination polymer $\{[Ni^{II}(pyrazole)_4]_2[Nb^{IV}(CN)_8]\cdot 4H_2O\}_n$ (Ni$_2$Nb) based on nickel(II) ($S = 1$) and niobium(IV) ($S = \frac{1}{2}$). We also discuss its properties in comparison to Mn^{II} and Fe^{II} analogs: $\{[Mn^{II}(pyrazole)_4]_2[Nb^{IV}(CN)_8]\cdot 4H_2O\}_n$ (Mn$_2$Nb) and $\{[Fe^{II}(pyrazole)_4]_2[Nb^{IV}(CN)_8]\cdot 4H_2O\}_n$ (Fe$_2$Nb) [13].

2. Results and Discussion

2.1. X-ray Crystal Structure Description under High Pressure

Ni$_2$Nb shows some interesting structural distortions when pressurized using a Merrill–Bassett DAC [14]. Its structure under pressure was determined by scXRD (Figure 1 and Table 1) [13]. Ni$_2$Nb crystallizes in a tetragonal space group $I4_1/a$ and forms a three-dimensional (3-D) CN-bridged skeleton with a flattened diamond-like topology where the niobium and nickel ions are all linked by cyanide ligands. The niobium(IV) centers play the role of the four-fold tetrahedral nodes in the 3-D framework of Ni$_2$Nb (Figure 1a,b).

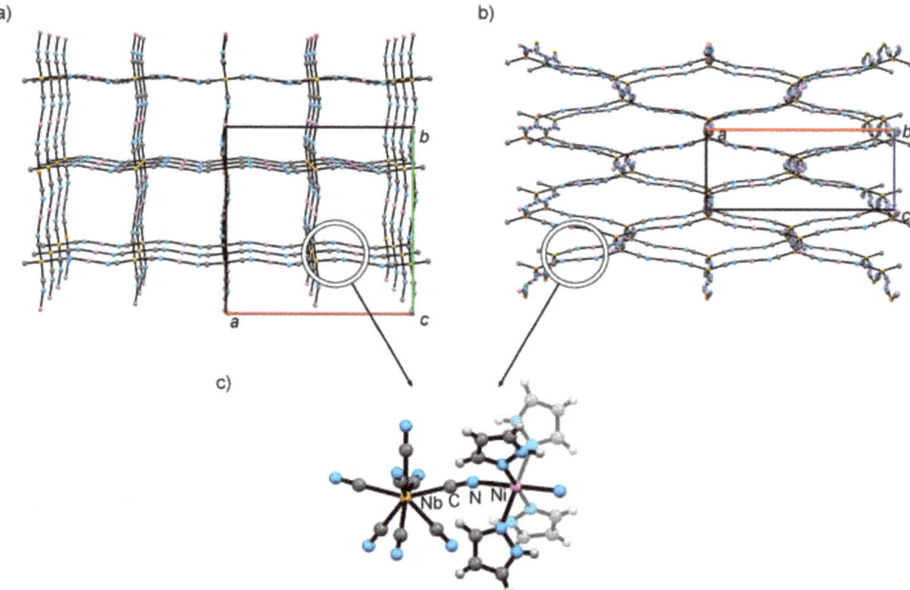

Figure 1. Structural diagrams presenting the cyanido-bridged framework of $\{[Ni^{II}(pyrazole)_4]_2[Nb^{IV}(CN)_8]\cdot 4H_2O\}_n$ down crystallographic axis z (**a**) and axis x (**b**). H$_2$O, pyrazole and non-bridging CN$^-$ are omitted for clarity. (**c**) The local geometry of the Nb^{IV}-CN-Ni^{II} structural motif.

Table 1. Single-crystal X-ray diffraction (scXRD) unit-cell parameters for Ni_2Nb at room temperature and high pressure.

Formula	$C_{32}H_{40}N_{24}NbNi_2O_4$
Temperature, K	296(2)
λ, Å	0.71073 Å
Molecular weight, g/mol	1035.11
Crystallographic system	tetragonal
Space group	$I4_1/a$
0.0001 GPa	
Unit cell, Å	$a = 21.4340(4)$
	$c = 9.6410(2)$
Volume V, Å^3	$V = 4429.23(15)$
0.25(2) GPa	
Unit cell, Å	$a = 21.2935(17)$
	$c = 9.584(2)$
Volume V, Å^3	$V = 4345.0(10)$
0.61(2) GPa	
Unit cell, Å	$a = 21.2002(7)$
	$c = 9.4404(5)$
Volume V, Å^3	$V = 4243.0(3)$
1.00(2) GPa	
Unit cell, Å	$a = 21.0921(13)$
	$c = 9.3306(11)$
Volume V, Å^3	$V = 4151.0(6)$
1.30(2) GPa	
Unit cell, Å	$a = 20.9721(8)$
	$c = 9.292(2)$
Volume V, Å^3	$V = 4087.1(1)$
1.50(2) GPa	
Unit cell, Å	$a = 20.8535(18)$
	$c = 9.219(7)$
Volume V, Å^3	$V = 4009(3)$
1.88(2) GPa	
Unit cell, Å	$a = 20.8438(9)$
	$c = 9.184(4)$
Volume V, Å^3	$V = 3990.3(17)$
2.15(2) GPa	
Unit cell, Å	$a = 20.704(10)$
	$c = 9.167(7)$
Volume V, Å^3	$V = 3930(4)$
2.48(2) GPa	
Unit cell, Å	$a = 20.635(16)$
	$c = 9.121(6)$
Volume V, Å^3	$V = 3884(5)$

The high-pressure compression of Ni_2Nb is presented in Table 1 and Figure 2 as the pressure dependence of the normalized unit-cell volume V/V_0, where V_0 is the unit-cell volume at 1000 hPa, along with the relevant data for the two analogs Mn_2Nb and Fe_2Nb published previously [8]. The unit-cell volumes of Mn_2Nb, Fe_2Nb and Ni_2Nb are significantly compressed up to 87.6%, 84.6% and

87.7% of the initial value at ca. 2.4 GPa, respectively. The V/V_0 vs. p dependences were fitted using the third-order Birch-Murnaghan equation of state (BMEOS) [15,16] (Equation (1)):

$$p(V) = \frac{3K_0}{2}\left[\left(\frac{V_0}{V}\right)^{\frac{7}{3}} - \left(\frac{V_0}{V}\right)^{\frac{5}{3}}\right]\left\{1 + \frac{3}{4}\left(K'_0 - 4\right)\left[\left(\frac{V_0}{V}\right)^{\frac{2}{3}} - 1\right]\right\} \tag{1}$$

where p—pressure, V_0—volume at zero pressure, in this case the ambient pressure, V—volume, K_0—isothermal bulk modulus at zero pressure, K'_0—dimensionless first derivative of K_0 with respect to pressure. The solid lines in Figure 2, which represent the best fit to Equation (1), match the experimental data for Ni_2Nb and Mn_2Nb in the investigated pressure range. In the case of Fe_2Nb there is a strong deviation from the BMEOS above 1 GPa. The best fit parameters K_0 and K_0' are: 10.7 ± 0.9 and 8.7 ± 1.8 GPa for Mn_2Nb, 10.4 ± 1.6 and 8.2 ± 5.6 GPa for Fe_2Nb (from the 0–1 GPa range fit) and 11.9 ± 1.2 and 6.9 ± 2.0 GPa in the case of Ni_2Nb. The bulk modulus K_0 value is identical for all isomorphs within the experimental error and quite similar to other molecule-based coordination compounds [6,17,18]. Noteworthy, the K_0 for the studied coordination polymers are nearly two orders of magnitude smaller than for diamond (440 GPa) and only one order larger than for rubber (1 GPa) [19]. This places the mechanical properties of coordination polymers somewhere between typical inorganic solids and soft matter.

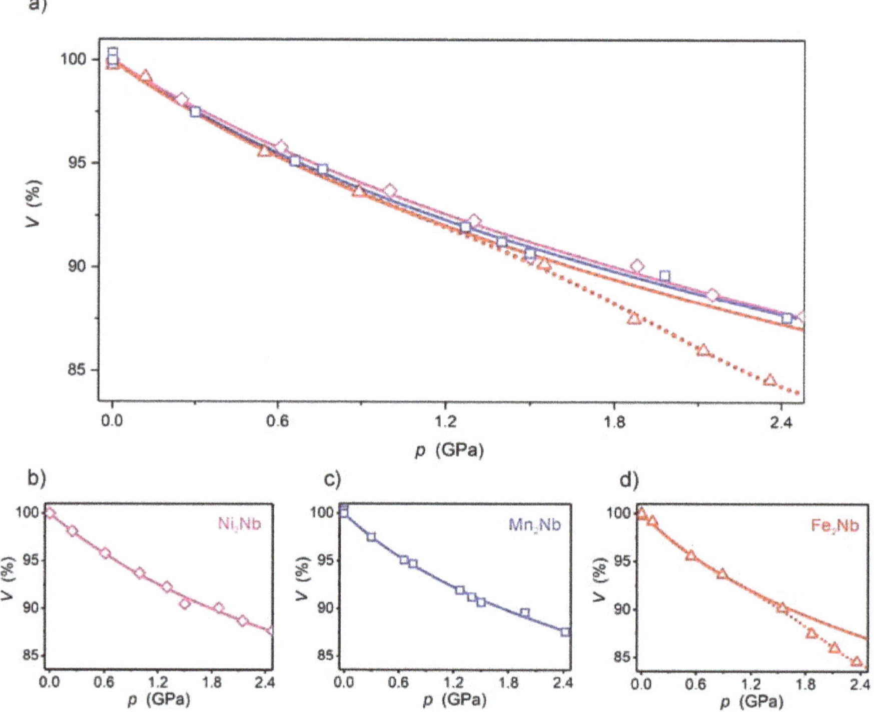

Figure 2. Combined graph (**a**) of $V/V_0(p)$ dependencies for Ni_2Nb (magenta) (**b**), Mn_2Nb (blue) (**c**) and Fe_2Nb (red) (**d**). The solid lines are the best fit to the Birch–Murnaghan equation of state (BMEOS). The dotted line for Fe_2Nb above 1.2 GPa is only a guide for the eye.

The analysis of the coordination spheres of Ni^{II} and Nb^{IV} in Ni_2Nb under high pressure leads to similar observations as for Mn_2Nb [8]: the Nb–C and C–N distances as well as Nb–C–N angles remain roughly unchanged, while the Ni–N bonds (Figure 3a) shrink significantly in a linear fashion

(Figure 3b). This fact and the good match between the $V/V_0(p)$ dependence and the BMEOS both indicate that no pressure-induced phase transition occurs at room temperature in Ni_2Nb and Mn_2Nb. In fact, the behavior of these two solids is very similar while that of Fe_2Nb is quite different and strongly deviates from BMEOS due to the SCO (SCO - spin crossover) behavior. Also, the Fe–N distances shrink in a non-linear fashion in Fe_2Nb above 1.5 GPa.

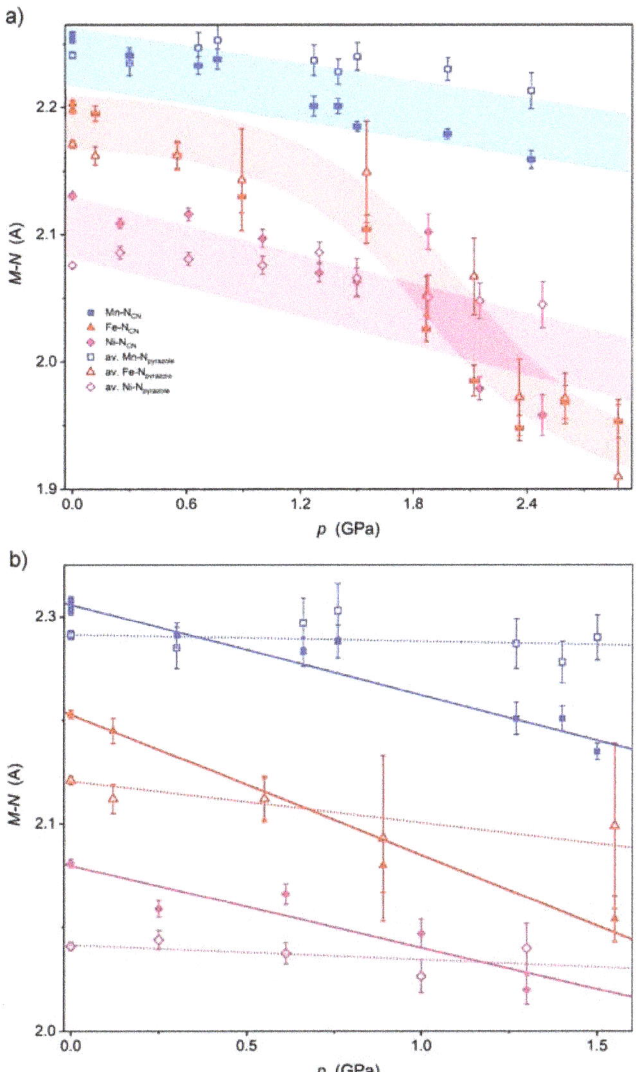

Figure 3. Pressure dependence of the $M–N_{CN}$ bond lengths (full symbols) and $M–N_{pyrazole}$ (open symbols) in the full pressure range (**a**) and in the pressure range where only linear changes are observed (**b**) for Ni_2Nb (magenta), Mn_2Nb (blue) and Fe_2Nb (red). Highlights in (**a**) are only for guiding the eye, while the lines in (**b**) are the best linear fit: solid for $M–N_{CN}$ and dotted for $M–N_{pyrazole}$.

A more detailed analysis of M–N bonds up to 1.5 GPa (Ni_2Nb, Mn_2Nb and Fe_2Nb follow the BMEOS in this pressure range) demonstrates that M–N_{CN} shrinkage is much larger than that of M–$N_{pyrazole}$ (Figure 3b and Table 2). As a result, the elongated octahedral coordination spheres of Fe^{II} in Fe_2Nb and Ni^{II} in Ni_2Nb become closer to a perfect octahedron at around 0.7 and 1.2 GPa, respectively.

Table 2. Shrinkage of the M–N_{CN} and M–$N_{pyrazole}$ bonds under pressure. The Δ(M–N)/Δp values are the slopes of the best linear fit from Figure 3b.

	Mn_2Nb	Fe_2Nb	Ni_2Nb
Δ(M-N_{CN})/Δp (Å GPa^{-1})	−0.044	−0.067	−0.040
Δ(M-$N_{pyrazole}$)/Δp (Å GPa^{-1})	−0.003	−0.020	−0.007

2.2. Magnetic Properties under High Pressure

The magnetic properties of Mn_2Nb and Fe_2Nb (Figures 4 and 5, respectively) have been discussed in detail in a previous report [8]. It was established that at ambient pressure both Mn_2Nb and Fe_2Nb are ferrimagnets with long-range magnetic ordering temperatures (T_c) of 23.4 and 9.4 K, respectively. However, the behavior of these compounds under high pressure is very different. Mn_2Nb (Figure 4) displays a very strong and linear increase of the T_c from 23.4 K at ambient pressure to 36.5 K at 1.03 GPa. The linear fit of the $T_c(p)$ dependence (Figure 4c) leads to $dT_c/dp = 12.4 \pm 0.2$ K GPa^{-1}, typical for octacyanidoniobate(IV)-based systems and Prussian Blue analogs [2]. Fe_2Nb, on the other hand, exhibits almost complete quenching of the long-range magnetic ordering under pressure, resulting in paramagnetic properties above 0.7 GPa (Figure 5).

Ni_2Nb is a ferromagnet due to the local ferromagnetic interactions between Nb^{IV} and Ni^{II} ions [13] with the Curie temperature (T_C) of 13.2 K. It exhibits a completely different type of magnetic response to high pressure (Figure 6) when compared to the other two analogs Mn_2Nb and Fe_2Nb. The T_C of Ni_2Nb decreases with increasing pressure, which is characteristic for ferromagnets [20,21], and confirms the presence of local ferromagnetic interactions between Nb^{IV} and Ni^{II}. The T_C shift is 1.9 K GPa^{-1}, as obtained from the linear fit of the $T_C(p)$ dependence (Figure 6c, T_C is the position of the $d\chi/dT$ peak in the $\chi(T)$ measurement at 10 Oe). The pressure response of Ni_2Nb is much weaker and opposite to Mn_2Nb, where the antiferromagnetic interactions between Nb^{IV} and Mn^{II} are present. Hence, the underlying local magnetic interactions in Ni_2Nb (ferromagnetic) vs. Mn_2Nb (antiferromagnetic) are the source of the observed difference. The shortening of the Ni–N_{CN} bonds favors the resonance integral and makes the antiferromagnetic contribution to the total exchange interaction stronger while destabilizing the ferromagnetic one. Overall, the J_{NiNb} coupling constant decreases under pressure.

Ni_2Nb does not present a loss of the magnetic moment under high pressure, which is confirmed by the pressure-independent magnetization at saturation values of 5.3 Nβ (at 2.0 K and 7 T; Figure 6b) up to 1.10 GPa and the structural pressure response that matches the BMEOS. The magnetization at saturation values are close to 5.2 Nβ, as expected for two Ni^{II} ($S = 1$) and one Nb^{IV} ($S = \frac{1}{2}$) coupled ferromagnetically (assuming $g_{Nb} = 2.0$ and $g_{Ni} = 2.1$ [22]).

Ni_2Nb also shows interesting pressure-induced changes in the $M(H)$ dependencies in the 0.1–5 T magnetic field range (Figure 6b). Following the initial sharp increase of the magnetization around 0.05 T in each case, there is a clear dependence with $M(H)$ attaining lower values at higher pressure. All $M(H)$ curves "meet" again above 5 T, converging to the same saturation value of 5.3 Nβ. This behavior is most probably related to the changes of the magnetic anisotropy of the Ni^{II} centers arising from the shortening of the Ni–N_{CN} bonds along the CN–Ni–CN axis of the Ni^{II} coordination sphere, as evidenced by high-pressure structural studies (Figure 3b). The contribution of $[Nb^{IV}(CN)_8]^{4-}$ to the pressure-induced magnetic anisotropy change in the M_2Nb family can be excluded based on the fact that $M(H)$ for Mn_2Nb does not change at all with increasing pressure (Figure 4b).

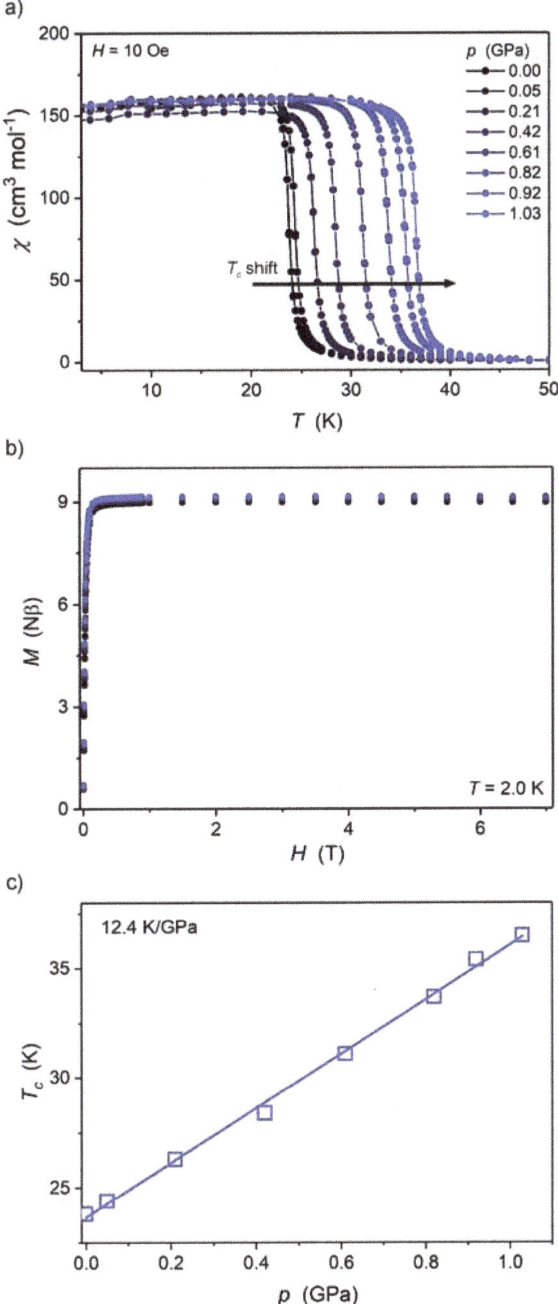

Figure 4. Temperature dependence of molar magnetic susceptibility $\chi(T)$ at 10 Oe (**a**), $M(H)$ at 2.0 K (**b**) and $T_c(p)$ (**c**) for Mn_2Nb under high pressure. The solid line in (**a**) is only a guide for the eye, while in (**c**) it represents the best linear fit.

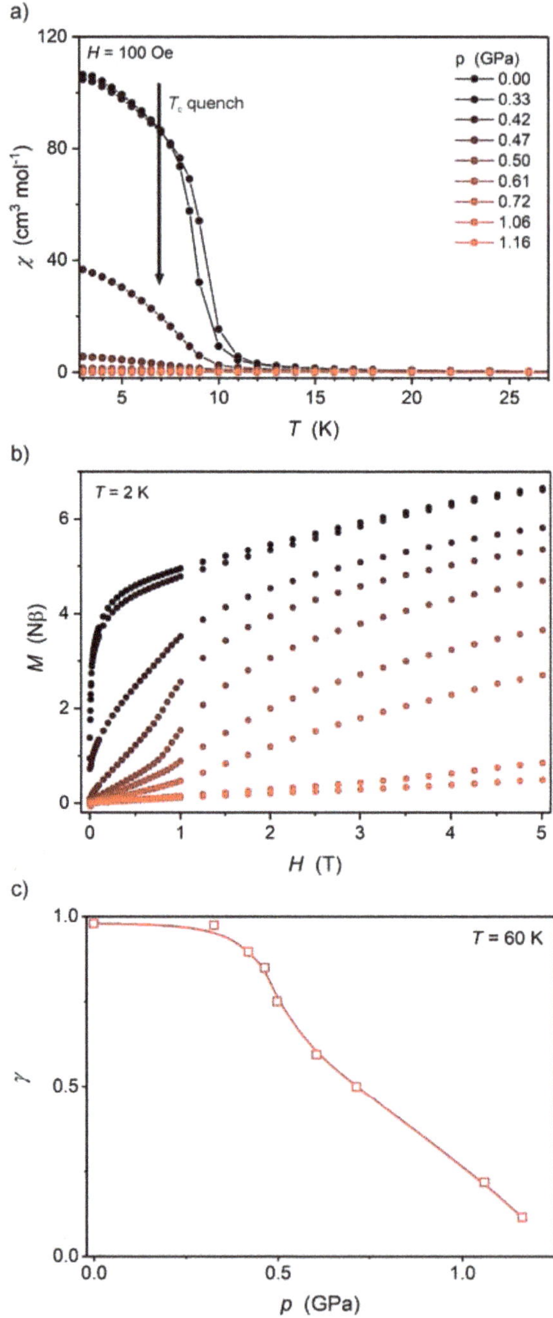

Figure 5. Temperature dependence of molar magnetic susceptibility $\chi(T)$ at 100 Oe (**a**), $M(H)$ at 2.0 K (**b**) and fraction of the high-spin Fe^{II} $\gamma(p)$ at 60 K (**c**) for Fe$_2$Nb under high pressure. The solid lines in (**a,c**) are only guides for the eye.

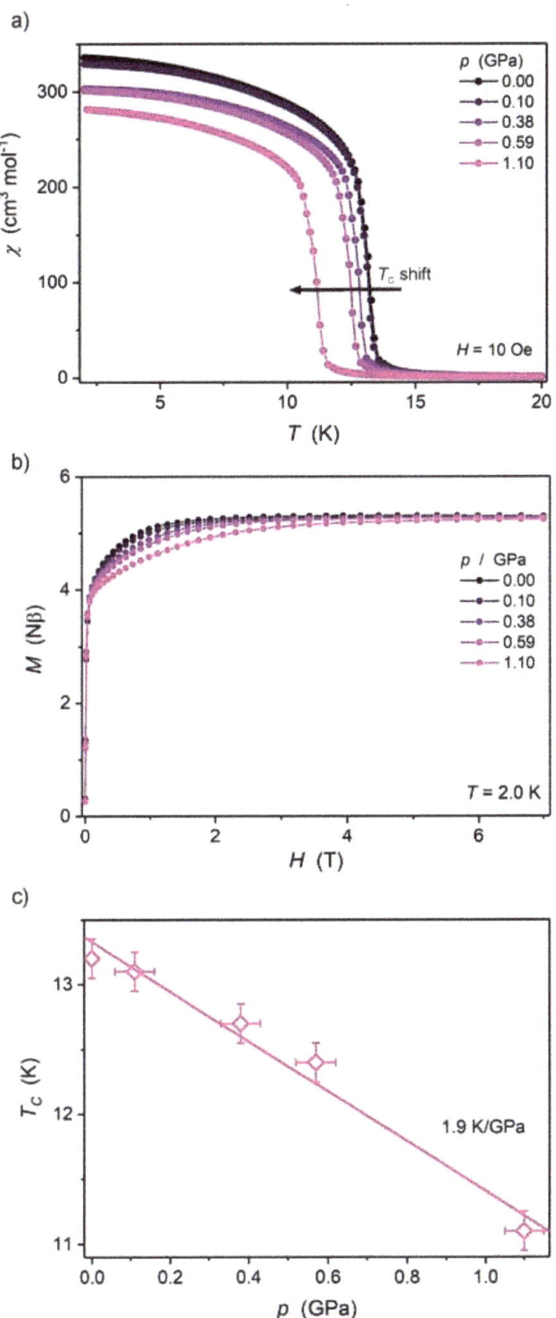

Figure 6. Temperature dependence of molar magnetic susceptibility $\chi(T)$ at 10 Oe (**a**), $M(H)$ at 2.0 K (**b**) and $T_c(p)$ (**c**) for Ni_2Nb under high pressure. The solid lines in (**a**,**b**) are only guides for the eye, while in (**c**) it represents the best linear fit.

3. Materials and Methods

3.1. Materials

Chemicals used in this study were of analytical grade and were obtained from commercial sources (Sigma-Aldrich Co., Avantor, Alfa-Aesar). $K_4[Nb(CN)_8] \cdot 2H_2O$ was prepared according to the newest available procedure [23]. $\{[Ni^{II}(pyrazole)_4]_2[Nb^{IV}(CN)_8] \cdot 4H_2O\}_n$ (Ni$_2$Nb) was obtained according to the literature procedure and its purity/identity was confirmed by elemental analysis and powder X-ray diffraction, which were identical to those published previously [13].

3.2. Single-Crystal X-ray Diffraction under Pressure

The single crystal diffraction data for Ni$_2$Nb were collected at room temperature for a single crystal placed in a Merrill–Bassett DAC filled with Fluorinert 77 as the pressure transmitting medium and a chip of ruby. The ruby fluorescence method was used for the pressure determination inside the DAC chamber. The diffraction frames were collected on KUMA KM4-CCD and Xcalibur EOS machines (running Crysalis software; instruments presently manufactured by Rigaku Oxford Diffraction) using a Mo Kα radiation source and a graphite monochromator ($\lambda = 0.71073$ Å). The data were corrected for the sample and DAC absorption as well as the for the gasket shadow [24]. Overlaps with the diamond reflections were excluded from the refinement. Non-H atoms were refined anisotropically (weighted full-matrix least-squares on F^2) [25]. The summary of crystallographic data can be found in Table 1. The Cambridge Crystallographic Data Center (CCDC) 1909715–1909722 contains the detailed supplementary crystallographic information for this paper. The crystallographic information files (CIFs) can be obtained free of charge from the CCDC via ww.ccdc.cam.ac.uk/data_request/cif.

The details of the related single-crystal XRD measurements under pressure for Mn$_2$Nb and Fe$_2$Nb were reported previously [8].

3.3. Magnetic Measurements under Pressure

Magnetic measurements under pressure for Mn$_2$Nb and Fe$_2$Nb were reported previously [8]. Ni$_2$Nb was characterized in a similar fashion using a Quantum Design MPMS3 SQUID-VSM (Sand Diego, USA) magnetometer. A powdered sample of Ni$_2$Nb was loaded into the CuBe piston-cylinder cell (manufactured by HMD, Japan; purchased from Quantum Design) with a piece of high-purity lead as the manometer and Daphne 7373 oil as the pressure-transmitting medium. The pressure determination at low temperature was performed with 0.02 GPa accuracy by using the linear pressure dependence of the superconducting transition of Pb (-0.379 K GPa^{-1}). Magnetic data were corrected for the diamagnetic contribution of the sample and the pressure cell.

4. Conclusions

A combined high-pressure magneto-structural study of a magnetic coordination polymer $\{[Ni^{II}(pyrazole)_4]_2[Nb^{IV}(CN)_8] \cdot 4H_2O\}_n$ Ni$_2$Nb was carried out and revealed a strong magnetic response of this material to mechanical stress. The magnetic ordering temperature of Ni$_2$Nb shifted linearly towards lower temperatures at higher pressure due to the ferromagnetic character of the exchange coupling between Nb^{IV} and Ni^{II} ions. Such behavior is completely different from that observed for the two analogs $\{[Mn^{II}(pyrazole)_4]_2[Nb^{IV}(CN)_8] \cdot 4H_2O\}_n$ Mn$_2$Nb and $\{[Fe^{II}(pyrazole)_4]_2[Nb^{IV}(CN)_8] \cdot 4H_2O\}_n$ Fe$_2$Nb. Mn$_2$Nb exhibited an opposite effect—a strong enhancement of T_c under pressure due to the antiferromagnetic exchange coupling between the constituent magnetic ions. Our study confirms the usefulness of combined high-pressure studies to understand the magnetic properties of molecular magnets and the possibility to fine-tune their properties by applying a mechanical stimulus—namely, high pressure.

Author Contributions: D.P. conceived and designed the experiments, performed the magnetic measurements and wrote the first draft of the paper; G.H. participated in the high-pressure magnetic measurements; H.T. and A.K. performed the high pressure scXRD measurements and analyzed the data; all authors participated in the preparation of the final version of the manuscript.

Funding: This research was funded by the Polish National Science Centre within the Sonata Bis project (2016/22/E/ST5/00055).

Conflicts of Interest: The authors declare no conflict of interest.

References

1. Woodall, C.H.; Craig, G.A.; Prescimone, A.; Misek, M.; Cano, J.; Faus, J.; Probert, M.R.; Parsons, S.; Moggach, S.; Martínez-Lillo, J.; et al. Pressure induced enhancement of the magnetic ordering temperature in rhenium(IV) monomers. *Nat. Commun.* **2016**, *7*, 13870. [CrossRef] [PubMed]

2. Pinkowicz, D.; Kurpiewska, K.; Lewiński, K.; Bałanda, M.; Mihalik, M.; Sieklucka, M.Z.B. High-pressure single-crystal XRD and magnetic study of a octacyanoniobate-based magnetic sponge. *CrystEngComm* **2012**, *14*, 5224–5229. [CrossRef]

3. Coronado, E.; Giménez-López, M.C.; Korzeniak, T.; Levchenko, G.; Romero, F.M.; Segura, A.; Garcia-Baonza, V.; Cezar, J.C.; de Groot, F.M.; Milner, A.; et al. Pressure-Induced Magnetic Switching and Linkage Isomerism in K0.4Fe4[Cr(CN)6]2.8·16H2O: X-ray Absorption and Magnetic Circular Dichroism Studies. *J. Am. Chem. Soc.* **2008**, *130*, 15519–15532. [CrossRef] [PubMed]

4. Ohba, M.; Kaneko, W.; Kitagawa, S.; Maeda, T.; Mito, M. Pressure Response of Three-Dimensional Cyanide-Bridged Bimetallic Magnets. *J. Am. Chem. Soc.* **2008**, *130*, 4475–4484. [CrossRef] [PubMed]

5. Shum, W.W.; Her, J.H.; Stephens, P.W.; Lee, Y.; Miller, J.S. Observation of the Pressure Dependent Reversible Enhancement of Tc and Loss of the Anomalous Constricted Hysteresis for [Ru2(O2CMe)4]3[Cr(CN)6]. *Adv. Mater.* **2007**, *19*, 2910–2913. [CrossRef]

6. Craig, G.A.; Sarkar, A.; Woodall, C.H.; Hay, M.A.; Marriott, K.E.; Kamenev, K.V.; Moggach, S.A.; Brechin, E.K.; Parsons, S.; Rajaraman, G.; et al. Rajaraman and M. Murrie. Probing the origin of the giant magnetic anisotropy in trigonal bipyramidal Ni(ii) under high pressure. *Chem. Sci.* **2018**, *9*, 1551–1559. [CrossRef]

7. Prescimone, A.; Sanchez-Benitez, J.; Kamenev, K.K.; Moggach, S.A.; Warren, J.E.; Lennie, A.R.; Murrie, M.; Parsons, S.; Brechin, E.K. High pressure studies of hydroxo-bridged Cu(ii) dimers. *Dalton Trans.* **2010**, *39*, 113–123. [CrossRef]

8. Pinkowicz, D.; Rams, M.; Mišek, M.; Kamenev, K.V.; Tomkowiak, H.; Katrusiak, A.; Sieklucka, B. Enforcing Multifunctionality: A Pressure-Induced Spin-Crossover Photomagnet. *J. Am. Chem. Soc.* **2015**, *137*, 8795–8802. [CrossRef]

9. Gütlich, P.; Ksenofontov, V.; Gaspar, A.B. Pressure effect studies on spin crossover systems. *Coord. Chem. Rev.* **2005**, *249*, 1811–1829. [CrossRef]

10. Katrusiak, A. High-pressure devices. In *International Tables for Crystallography, Volume H, Powder Diffraction*; Gilmore, C.J., Kaduk, J.A., Schenk, H., Eds.; John Wiley & Sons: New York, NY, USA, 2018; pp. 156–173. [CrossRef]

11. Kamarád, J.; Machátová, Z.; Arnold, Z. High pressure cells for magnetic measurements—Destruction and functional tests. *Rev. Sci. Instrum.* **2004**, *75*, 5022–5025. [CrossRef]

12. Awaga, K.; Sekine, T.; Okawa, M.; Fujita, W.; Holmes, S.M.; Girolami, G.S. High-pressure effects on a manganese hexacyanomanganate ferrimagnet with TN = 29K. *Chem. Phys. Lett.* **1998**, *293*, 352–356. [CrossRef]

13. Pinkowicz, D.; Pełka, R.; Drath, O.; Nitek, W.; Bałanda, M.; Majcher, A.M.; Poneti, G.; Sieklucka, B. Nature of Magnetic Interactions in 3D {[MII(pyrazole)4]2[NbIV(CN)8]·4H2O}n (M = Mn, Fe, Co, Ni) Molecular Magnets. *Inorg. Chem.* **2010**, *49*, 7565–7576. [CrossRef]

14. Merrill, L.; Bassett, W.A. Miniature diamond anvil pressure cell for single crystal x-ray diffraction studies. *Rev. Sci. Instrum.* **1974**, *45*, 290–294. [CrossRef]

15. Birch, F. Finite Elastic Strain of Cubic Crystals. *Phys. Rev.* **1947**, *71*, 809–824. [CrossRef]

16. Birch, F. Finite strain isotherm and velocities for single-crystal and polycrystalline NaCl at high pressures and 300 K. *J. Geophys. Res. Solid Earth* **1978**, *83*, 1257–1268. [CrossRef]

Magnetochemistry **2019**, *5*, 33

17. Byrne, P.J.; Richardson, P.J.; Chang, J.; Kusmartseva, A.F.; Allan, D.R.; Jones, A.C.; Kamenev, K.V.; Tasker, P.A.; Parsons, S. Piezochromism in Nickel Salicylaldoximato Complexes: Tuning Crystal-Field Splitting with High Pressure. *Chem. Eur. J.* **2012**, *18*, 7738–7748. [CrossRef] [PubMed]

18. Madsen, S.R.; Overgaard, J.; Stalke, D.; Iversen, B.B. High-pressure single crystal X-ray diffraction study of the linear metal chain compound $Co_3(dpa)_4Br_2 \cdot CH_2Cl_2$. *Dalton Trans.* **2015**, *44*, 9038–9043. [CrossRef] [PubMed]

19. Tabor, D. The bulk modulus of rubber. *Polymer* **1994**, *35*, 2759–2763. [CrossRef]

20. Motokawa, N.; Miyasaka, H.; Yamashita, M. Pressure effect on the three-dimensional charge-transfer ferromagnet [{Ru_2(m-FPhCO$_2$)$_4$}$_2$(BTDA-TCNQ)]. *Dalton Trans.* **2010**, *39*, 4724–4726. [CrossRef] [PubMed]

21. Mito, M.; Matsumoto, K.; Komorida, Y.; Deguchi, H.; Takagi, S.; Tajiri, T.; Iwamoto, T.; Kawae, T.; Tokita, M.; Takeda, K. Volume shrinkage dependence of ferromagnetic moment in lanthanide ferromagnets gadolinium, terbium, dysprosium, and holmium. *J. Phys. Chem. Solids* **2009**, *70*, 1290–1296. [CrossRef]

22. Kataev, V.; Golze, C.; Alfonsov, A.; Klingeler, R.; Büchner, B.; Goiran, M.; Broto, J.M.; Rakoto, H.; Mennerich, C.; Klauss, H.H.; et al. Magnetism of a novel tetranuclear nickel(II) cluster in strong magnetic fields. *J. Phys. Conf. Ser.* **2006**, *51*, 351–354. [CrossRef]

23. Handzlik, G.; Magott, M.; Sieklucka, B.; Pinkowicz, D. Alternative Synthetic Route to Potassium Octacyanidoniobate (IV) and Its Molybdenum Congener. *Eur. J. Inorg. Chem.* **2016**, *2016*, 4872–4877. [CrossRef]

24. Katrusiak, A. Shadowing and absorption corrections of single-crystal high-pressure data. *Zeitschrift Für Kristallographie Crystalline Materials* **2004**, *219*, 461–467. [CrossRef]

25. Sheldrick, G. A short history of SHELX. *Acta Crystallogr. Sect. A Found. Crystallogr.* **2008**, *64*, 112–122. [CrossRef] [PubMed]

 *magnetochemistry*

Article

Chloranilato-Based Layered Ferrimagnets with Solvent-Dependent Ordering Temperatures

Cristian Martínez-Hernández, Samia Benmansour * and Carlos J. Gómez-García *

Instituto de Ciencia Molecular (ICMol), Departamento de Química Inorgánica, Universidad de Valencia, C/Catedrático José Beltrán 2, 46980 Paterna, Spain; christian.martinez@uv.es
* Correspondence: sam.ben@uv.es (S.B.); carlos.gomez@uv.es (C.J.G.-G.);
   Tel.: +34-963-544-423 (S.B. & C.J.G.-G.); Fax: +34-963-543-273 (S.B. & C.J.G.-G.)

Received: 2 May 2019; Accepted: 28 May 2019; Published: 4 June 2019

Abstract: We report the synthesis and the characterization of six new heterometallic chloranilato-based ferrimagnets formulated as $(NBu_4)[MnCr(C_6O_4Cl_2)_3]\cdot nG$ with n = 1 for G = C_6H_5Cl (**1**), C_6H_5I (**3**), and $C_6H_5CH_3$ (**4**); n = 1.5 for G = C_6H_5Br (**2**) and n = 2 for G = C_6H_5CN (**5**) and $C_6H_5NO_2$ (**6**); $(C_6O_4Cl_2)^{2-}$ = 1,3-dichloro,2,5-dihydroxy-1,4-benzoquinone dianion. The six compounds are isostructural and show hexagonal honeycomb layers of the type $[MnCr(C_6O_4Cl_2)_3]^-$ alternating with layers containing the NBu_4^+ cations. The hexagons are formed by alternating Mn(II) and Cr(III) connected by bridging bis-bidentate chloranilato ligands. The benzene derivative solvent molecules are located in the hexagonal channels (formed by the eclipsed packing of the honeycomb layers) showing π-π interactions with the anilato rings. The six compounds behave as ferrimagnets with ordering temperatures in the range 9.8–11.2 K that can be finely tuned by the donor character of the benzene ring and by the number of solvent molecules inserted in the hexagonal channels. The larger the electron density on the aromatic ring and the larger the number of solvent molecules are, the higher T_c is. The only exception is provided by toluene, where the formation of H-bonds might be at the origin of weaker π-π interactions observed in this compound.

Keywords: two-dimensional (2D) ferrimagnets; chloranilato; heterometallic layers; honeycomb layers; molecule-based magnets

---

1. Introduction

One of the main advantages of molecule-based magnets is the possibility to modulate or tune the properties of the magnets by simply changing or modifying the building blocks used to prepare them [1]. This strategy led, at the end of last century, to the synthesis of different series of molecule-based magnets whose ordering temperatures and coercive fields could be modified with ease. A typical example is provided by the series of cyano-bridged heterometallic compounds formulated as $A_xM_y[M'(CN)_6]_z\cdot nH_2O$, where A is a monovalent cation, and M and M' are trivalent or divalent transition metal ions [2–4]. In this series, the magnetic exchange through the CN bridge can be modulated [5,6] by changing A, M, and M' to obtain materials with interesting magnetic properties as photomagnetism [7–9], single molecule magnets [10,11], and even magnetic order above room temperature [12]. Another family of molecule-based magnets whose properties can be easily modified by changing the constituent metallic atoms is the series of oxalato-based two-dimensional (2D) magnets formulated as $(A)[M^{II}M^{III}(C_2O_4)_3]$ (A^+ = monocation; M^{II} = Mn, Fe, Co, Ni, Cu, … ; M^{III} = Fe, Cr, … ; $C_2O_4^{2-}$ = oxalate dianion, Figure 1b) that show ferro-, ferri-, or canted antiferromagnetic ordering with T_c ranging from 6 K to 48 K depending on M(II) and M(III) [13–22].

A third and recent example is the series of anilato-based heterometallic 2D honeycomb magnets formulated as $(A)[M^{II}M^{III}(C_6O_4X_2)_3]\cdot G$, where A^+ is a monocation (see Table 1); M(II) and M(III) are transition metal ions as Mn(II), Fe(II), Cr(III), and Fe(III), G may be many different solvent molecules (see Table 1), and $C_6O_4X_2^{2-}$ is the 1,3-disubstituted-2,5-dihydroxy- 1,4-benzoquinone dianion (with X = H, Cl, Br, NO_2, ... Figure 1a), known as anilato-type ligands [23]. This family of magnets shows a honeycomb (6,3)-2D structure with the same topology as the oxalato ones and, as in the oxalato family, it is also possible to change the magnetic properties by simply changing the building blocks [23,24]. Albeit, there are three important differences between these two series; the first difference is observed in the sign of the magnetic coupling—it is always antiferromagnetic in the anilato-based compounds, whereas it may be ferro- or antiferromagnetic depending on the metal ions, for the oxalato series. This fact precludes the presence of magnetic ordering in the homo-metallic anilato-based lattices but not in the hetero-metallic ones, where long range ferrimagnetic is observed when the spin states of the metal ions are different [Mn(II)Cr(III) and Fe(II)Fe(III), see Table 1]. The second difference is the rigidity of the oxalato ligand (Figure 1b) in contrast to the anilato ligand that can be easily modified by changing the X group (X = H, F, Cl, Br, I, CH_3, Cl/CN, NO_2, ...) [25]. This change has already allowed a tuning of the ordering temperatures in the series of compounds $(NBu_4)[MnCr(C_6O_4X_2)_3]$ (X = H, Cl, Br, and I) [23] The third important difference is the size; the hexagonal cavities of the honeycomb structure are twice as large in the anilato-based compounds and, when packed in an eclipsed way, originate hexagonal channels with BET areas of up to 1440 m^2/g [26]. These hexagonal channels may be filled with solvent molecules (in contrast to the oxalato-based compounds) that can be easily removed, giving rise, in some cases, to important changes in the magnetic properties. Thus, in compound $(NMe_2H_2)_2[Fe_2(C_6O_4Cl_2)_3]\cdot2H_2O\cdot6DMF$, the removal of the solvent molecules results in a decrease of the ordering temperature from 80 to 26 K [27]. In compound $(Et(i-Pr)_2NH)[MnCr(C_6O_4Br_2)_3]\cdot H_2O\cdot0.5CHCl_3$, the removal of the solvent molecules changes the magnetic behavior (the solvated compound is a metamagnet with a critical field of 490 mT at 2 K, whereas the de-solvated phase is a ferrimagnet with T_c = 9 K) [28].

The possibility to change the magnetic properties (ordering temperatures, magnetic behavior, or critical and coercive fields) by simply changing the solvent molecules is, therefore, a very appealing strategy to modulate T_c in these series of 2D magnets. Furthermore, when using lanthanoids as metal ions, the solvent molecules also play a key structural role in these (6,3)-2D lattices [24,29–38].

In this context, we have recently initiated a detailed study of the role played by the solvent molecules located in the hexagonal channels in the structure and the magnetic properties of these 2D magnets formulated as $(A)[M^{II}M^{III}(C_6O_4X_2)_3]\cdot G$. To perform this study, we have initially selected Mn(II) and Cr(III) as M^{II} and M^{III}, since this couple of metal ions crystallizes more easily (Table 1). We have selected A = NBu_4^+ as the cation and chloranilato as the ligand (X = Cl), and we have focused on a series of benzene derivative solvent molecules (C_6H_5X with X = Cl, Br, I, CH_3, CN, and NO_2), since they seem to play a template role that facilitates the crystallization of these 2D lattices. With this idea in mind, we have prepared the series of compounds formulated as $(NBu_4)[MnCr(C_6O_4Cl_2)_3]\cdot n\ C_6H_5X$ with n/X = 1/Cl (**1**), 1.5/Br (**2**), 1/I (**3**), 1/CH_3 (**4**), 2/CN (**5**), and 2/NO_2 (**6**). This series of compounds are solvates since they are isostructural and only differ in the solvent molecules. They present the structure of compound $(NBu_4)[MnCr(C_6O_4Br_2)_3]\cdot C_6H_5Br\cdot0.5H_2O$ (**A**) [39] and show a fine modulation of the ordering temperatures in the range 9.8–11.2 K depending on the electronic properties of the benzene derivative molecule and on the number of solvent molecules inserted in the channels.

Here, we present the chemical and the magnetic characterization of the six compounds and show that it is possible to fine-tune the ordering temperatures with a simple change of the solvent molecules.

Figure 1. (**a**) The anilato family of ligands (C$_6$O$_4$X$_2$)$^{2-}$ showing the typical bis-bidentate coordination mode also shown by the oxalato ligand (**b**).

Table 1. Magnetic properties of all the structurally characterized hetero-metallic layered compounds of the type (A)[M^{II}M^{III}(C$_6$O$_4$X$_2$)$_3$]·(solvent).

CCDC	Formula	Packing	Space Group	T_c (K)	H_{coer} (mT) [a]	Ref.
MIRFEA	[(H$_3$O)(phz)$_3$][MnCr(C$_6$O$_4$Cl$_2$)$_3$(H$_2$O)] [b]	eclipsed	*P3*	5.5	19.4	[23]
MIRFIE	[(H$_3$O)(phz)$_3$][MnCr(C$_6$O$_4$Br$_2$)$_3$]·2H$_2$O·2CH$_3$COCH$_3$	eclipsed	*P-31m*	6.3	34.0	[23]
MIRFOK	[(H$_3$O)(phz)$_3$][MnFe(C$_6$O$_4$Br$_2$)$_3$]·H$_2$O	eclipsed	*P-31m*	-	-	[23]
MIRFUQ	(NBu$_4$)[MnCr(C$_6$O$_4$Cl$_2$)$_3$]	alternated	*C2/c*	5.5	11.8	[23]
HOWHAE	[Fe(sal$_2$-trien)][MnCr(C$_6$O$_4$Cl$_2$)$_3$]·0.5CH$_2$Cl$_2$·CH$_3$OH·0.5H$_2$O·5CH$_3$CN	alternated	*C222$_1$*	10.0	35	[40]
HOWHEI	[Fe(4-OH-sal$_2$-trien)][MnCr(C$_6$O$_4$Cl$_2$)$_3$]·G	alternated	*P6$_1$22*	10.4	87	[40]
HOWHIM	[Fe(sal$_2$-epe)][MnCr(C$_6$O$_4$Br$_2$)$_3$]·4CH$_3$CN	alternated	*P21/c*	10.2	10	[40]
HOWHOS	[Fe(5-Cl-sal$_2$-trien)][MnCr(C$_6$O$_4$Br$_2$)$_3$]·CH$_2$Cl$_2$·CH$_3$OH·4H$_2$O·1.5CH$_3$CN	alternated	*P21/c*	9.8	66	
MUMKUC	[Fe(acac$_2$-trien)][MnCr(C$_6$O$_4$Cl$_2$)$_3$]·2CH$_3$CN	alternated	*C2/c*	10.8	65	[41]
MUMLAJ	[Fe(acac$_2$-trien)][MnCr(C$_6$O$_4$Br$_2$)$_3$]·2CH$_3$CN	alternated	*C2/c*	11.1	77	[41]
MUMLEN	[Ga(acac$_2$-trien)][MnCr(C$_6$O$_4$Br$_2$)$_3$]·2CH$_3$CN	alternated	*C2/c*	11.6	72	[41]
SEPLAD	(Me$_2$NH$_2$)[MnCr(C$_6$O$_4$Br$_2$)$_3$]·2H$_2$O	alternated	*P-31c*	7.9	90	[28]
SEPLEH	(Et$_2$NH$_2$)[MnCr(C$_6$O$_4$Br$_2$)$_3$]	alternated	*P-31c*	8.9	100	[28]
SEQCID	(Et$_3$NH)[MnCr(C$_6$O$_4$Cl$_2$)$_3$]	alternated	*P-31c*	8.0	150	[28]
SEPROX	(Et(i-Pr)$_2$NH)[MnCr(C$_6$O$_4$Br$_2$)$_3$]	alternated	*P-31c*	9.0	4 [c]	[28]
1910770	(NBu$_4$)[MnCr(C$_6$O$_4$Br$_2$)$_3$]·C$_6$H$_5$Br·0.5H$_2$O	eclipsed	*P2$_1$*	9.5	33	[39]
QEFPOJ	[(H$_3$O)(phz)$_3$][FeFe(C$_6$O$_4$Cl$_2$)$_3$]·12H$_2$O [d]	eclipsed	*P-31m*	2.4	1.0	[42]
QEFPID	[(H$_3$O)(phz)$_3$][FeFe(C$_6$O$_4$Br$_2$)$_3$]·12H$_2$O [d]	eclipsed	*P-31m*	2.1	1.0	[42]
NIHJEW01	[C(N$_2$H$_3$)$_3$][FeFe(C$_6$O$_4$(CN)Cl)$_3$]·29H$_2$O [d]	eclipsed	*P3*	4.0	6	[43]
1909314	(NBu$_4$)[MnCr(C$_6$O$_4$Cl$_2$)$_3$(C$_6$H$_5$CHO)]·C$_6$H$_6$ [b]	eclipsed	*P2$_1$*	7.0	7.6	[44]
1909315	(NBu$_4$)[MnCr(C$_6$O$_4$Br$_2$)$_3$(C$_6$H$_5$CHO)]·C$_6$H$_6$ [b]	eclipsed	*P2$_1$*	6.7	10	[44]
1909316	(NBu$_4$)[MnCr(C$_6$O$_4$Cl$_2$)$_3$(C$_6$H$_5$CHO)]·C$_6$H$_5$CHO [b]	eclipsed	*P2$_1$*	6.8	5.0	[44]
1909317	(NBu$_4$)[MnCr(C$_6$O$_4$Br$_2$)$_3$(C$_6$H$_5$CHO)]·C$_6$H$_5$CHO [b]	eclipsed	*P2$_1$*	6.7	20	[44]

[a] at 2 K; [b] the H$_2$O or C$_6$H$_5$CHO molecule is coordinated to the Mn ion; [c] A solvated phase of this compound is metamagnetic with a critical field at 2 K of 490 mT. [d] These compounds are homo-metallic but show two different oxidation states.

2. Results and Discussion

2.1. Syntheses of the Complexes

The six compounds were synthesized by carefully layering solutions containing the precursor [Cr(C$_6$O$_4$Cl$_2$)$_3$]$^{3-}$ anion and Mn(II) ions with the corresponding benzene derivative solvents. In all cases, an intermediate layer was needed to slow down the crystallization process and prevent the formation of amorphous or low quality crystalline materials. All the attempts to obtain good quality single crystals failed, although the X-ray powder diffractograms showed that they are all isostructural to the closely related compound $(NBu_4)[MnCr(C_6O_4Br_2)_3] \cdot C_6H_5Br \cdot 0.5H_2O$ (**A**) [39].

2.2. FT-IR Spectra

As expected, the six compounds showed very similar IR spectra (Figure 2). The only differences corresponded to the bands associated with the solvent molecules. Table 2 lists the main bands and their assignments. The FT-IR spectra in all cases confirmed the presence of the corresponding solvent molecules (C_6H_5Cl in **1**, C_6H_5Br in **2**, C_6H_5I in **3**, $C_6H_5CH_3$ in **4**, C_6H_5CN in **5**, and $C_6H_5NO_2$ in **6**), in agreement with the chemical and the thermo-gravimetric analysis.

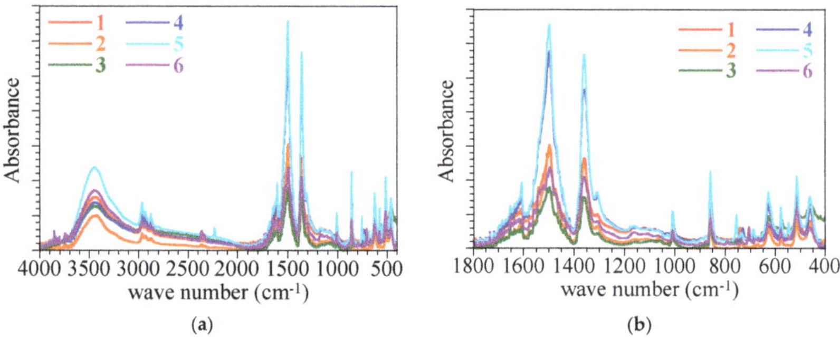

Figure 2. (a) Absorbance FT-IR spectra of compounds **1–6** in the range 4000–400 cm^{-1} (a) and 1800–400 cm^{-1} (b).

Table 2. Main IR bands (cm^{-1}) and their assignments for compounds **1–6**.

Band	1 (C_6H_5Cl)	2 (C_6H_5Br)	3 (C_6H_5I)	4 (C_6H_5CN)	5 ($C_6H_5CH_3$)	6 ($C_6H_5NO_2$)
ν(C-H)	2959 2929 2873	2959 2929 2873	2960 2928 2873	2962 2931 2873	2960 2928 2873	2963 2931 2873
ν(C≡N) [1]	-	-	-	2227	-	-
ν(C=O)	1609	1606	1607	1607	1607	1606
ν(N-O)$_{as}$ [1]	-	-	-	-	-	1520
ν(C=C) [1]	1506	1506 *	1505	1507 *	1505 *	1505
ν(C=C) + ν(C-O)	1495	1496	1496	1497	1496	1495
δ(C-H)	1383	1382	1383	1383 *	1383 *	1383
ν(C-C) + ν(C-O)	1360	1362	1361	1360	1359	1361
ν(N-O)$_s$ [1]	-	-	-	-	-	1344 *
ν(C-N) [1]	-	-	-	-	-	877
δ(C-X)	860	858	858	858	858	857
ν(C-Cl) [1]	740	-	-	-	-	-
δ(C-H) [1]	700 684	737 682	730 685	734 687	730 693	705 683
ρ(C-X)	577	576	576	577	576	577

* shoulder; [1] Bands corresponding to the solvent molecules.

2.3. Thermogravimetric Analysis

The thermogravimetric analysis of compounds **1–6** (Figure 3) showed an initial weight loss with a plateau in the range *ca.* 200–290 °C depending on the sample (Table 3). This weight loss corresponded to the release of the benzene derivative solvent molecules. The experimental weight loss (Table 3) indicated that compounds **1**, **3**, and **4** contained one solvent molecule (C_6H_5Cl in **1**, C_6H_5I in **3**, and $C_6H_5CH_3$ in **4**), whereas compound **2** contained 1.5 C_6H_5Br molecules and compounds **5** and **6** contained two solvent molecules (C_6H_5CN in **5** and $C_6H_5NO_2$ in **6**). These values are in agreement with the elemental analysis in all cases (see experimental section). We can, therefore, conclude that the used solvents (C_6H_5Cl, C_6H_5Br, C_6H_5I, $C_6H_5CH_3$, C_6H_5CN, and $C_6H_5NO_2$) entered in the structures of compounds **1–6** (as observed in the IR spectra), and that compounds **1**, **3**, and **4** contain one solvent molecule, compound **2** contains ca. 1.5, and compounds **5** and **6** contain around two solvent molecules per formula. At higher temperatures (around 350 °C), all compounds showed an abrupt weight loss corresponding to the decomposition and the release of the chloranilato ligand. As can be seen in the derivative plot, compound **4** needed a higher temperature (around 300 °C) to release the toluene molecule, suggesting that this molecule has a stronger interaction with the 2D lattice (probably due to the formation of H-bonds, as already observed in other anilato-based lattices) [34].

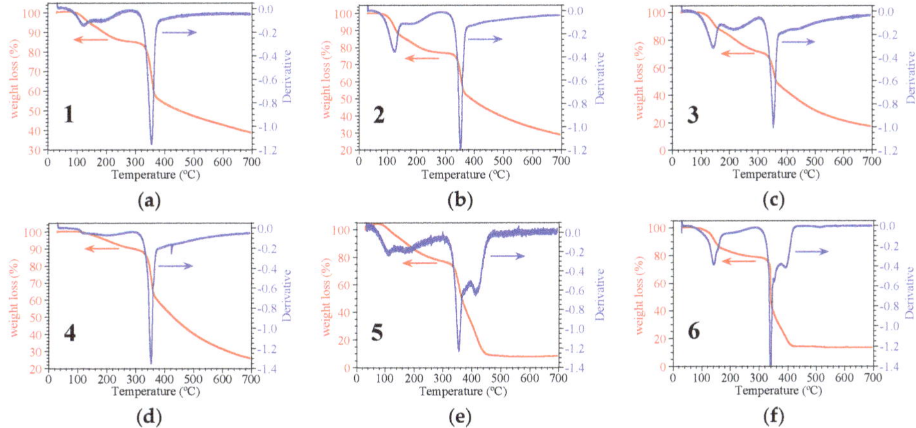

Figure 3. Thermogravimetric measurements and the corresponding derivative curves in the temperature range 30–700 °C for compounds **1** (a), **2** (b), **3** (c), **4** (d), **5** (e), and **6** (f).

Table 3. Experimental and calculated weigh losses (%) for compounds **1–6** at the first plateau at around 250–300 °C.

Compound	Temperature Range (°C)	Experimental Weight Loss (%)	Solvent	Calculated Weight Loss
$(NBu_4)[MnCr(C_6O_4Cl_2)_3]\cdot C_6H_5Cl$ (**1**)	30–200	10.8	1 C_6H_5Cl	10.4
$(NBu_4)[MnCr(C_6O_4Cl_2)_3]\cdot 1.5C_6H_5Br$ (**2**)	30–250	22.0	1.5 C_6H_5Br	19.5
$(NBu_4)[MnCr(C_6O_4Cl_2)_3]\cdot C_6H_5I$ (**3**)	30–200	16.6	1 C_6H_5I	17.4
$(NBu_4)[MnCr(C_6O_4Cl_2)_3]\cdot C_6H_5CH_3$ (**4**)	30–290	9.7	1 $C_6H_5CH_3$	8.7
$(NBu_4)[MnCr(C_6O_4Cl_2)_3]\cdot 2C_6H_5CN$ (**5**)	30–250	16.8	2 C_6H_5CN	17.5
$(NBu_4)[MnCr(C_6O_4Cl_2)_3]\cdot 2C_6H_5NO_2$ (**6**)	30–220	20.7	2 $C_6H_5NO_2$	20.2

2.4. Structures of Compounds **1–6**

Compounds **1–6** are isostructural to compound $(NBu_4)[MnCr(C_6O_4Br_2)_3]\cdot C_6H_5Br\cdot 0.5H_2O$ (**A**), [39] as shown by their X-ray powder diffractograms (Figure 4). The unit cell parameters of compounds **1–6** (Table 4), calculated from their X-ray powder diffractograms based on the structure of **A** with X'Pert HighScore Plus software, [45] further confirmed the isostructurality. Interestingly, compounds **1–6** did not show any correlation between the size and the number of the solvent molecules and the unit cell parameters, suggesting that the solvent molecules are located in the hexagonal channels rather than in the interlayer space and, therefore, they do not modify the structure.

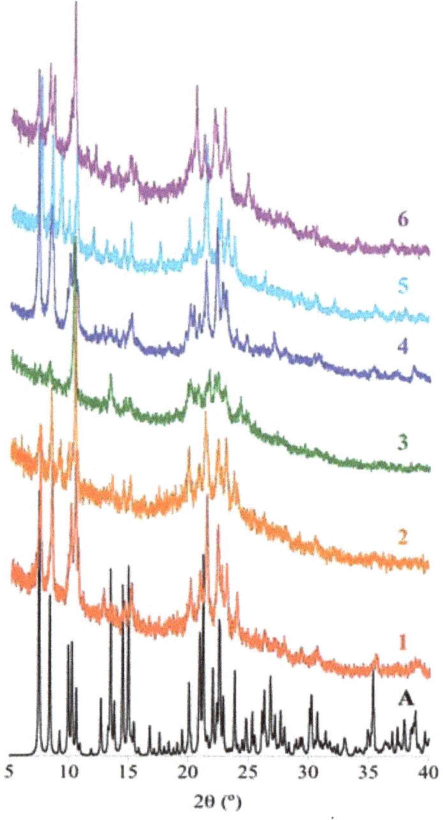

Figure 4. Simulated X-ray powder diffractogram of compound $(NBu_4)[MnCr(C_6O_4Br_2)_3]\cdot C_6H_5Br\cdot 0.5H_2O$ (**A**) and experimental diffractograms of compounds **1–6**.

Table 4. Unit cell parameters of compounds **1–6** determined from their X-ray powder diffractograms at room temperature.

Compound	a (Å)	b (Å)	c (Å)	(°)	Volume (Å³)
$(NBu_4)[MnCr(C_6O_4Br_2)_3]\cdot C_6H_5Br\cdot 0.5H_2O$ (**A**) [a]	9.9557(5)	23.6054(10)	12.2129(6)	105.187(5)	2769.9(2)
$(NBu_4)[MnCr(C_6O_4Br_2)_3]\cdot C_6H_5Br\cdot 0.5H_2O$ (**A**) [b]	10.104(4)	23.70(1)	12.363(5)	106.78(4)	2834.17
$(NBu_4)[MnCr(C_6O_4Cl_2)_3]\cdot C_6H_5Cl$ (**1**)	9.799(8)	23.76(3)	12.08(1)	104.71(8)	2719.83
$(NBu_4)[MnCr(C_6O_4Cl_2)_3]\cdot 1.5C_6H_5Br$ (**2**)	9.89(1)	23.66(4)	12.06(2)	103.7(1)	2740.18
$(NBu_4)[MnCr(C_6O_4Cl_2)_3]\cdot C_6H_5I$ (**3**)	9.71(4)	23.6(1)	12.23(6)	104.3(5)	2720.50
$(NBu_4)[MnCr(C_6O_4Cl_2)_3]\cdot C_6H_5CH_3$ (**4**)	9.822(7)	23.88(3)	12.09(1)	104.95(8)	2739.65
$(NBu_4)[MnCr(C_6O_4Cl_2)_3]\cdot 2C_6H_5CN$ (**5**)	9.887(9)	23.55(3)	12.08(1)	104.5(1)	2723.03
$(NBu_4)[MnCr(C_6O_4Cl_2)_3]\cdot 2C_6H_5NO_2$ (**6**)	9.58(4)	23.7(1)	11.98(6)	105.8(4)	2611.89

[a] Single crystal X-ray data at 120 K [39]. [b] Determined from X-ray powder data at room temperature.

Compound **A** (and compounds **1–6**) presents a layered structure where anionic layers of formula $[MnCr(C_6O_4X_2)_3]^-$ with X = Br (**A**) and Cl (**1–6**) alternate with cationic layers of NBu_4^+ cations (Figure 5a). The interlayer distance is 8.7 Å. The anionic layers show the classical honeycomb structure with Mn(II) and Cr(III) ions alternating in the vertex of the hexagons and the anilato ligands in the sides (Figure 5b). The benzene derivative solvent molecules are located in the hexagonal channels formed by the eclipsed packing of the honeycomb layers (Figure 5c) and show strong π-π interactions with the anilato rings with an interplane angle of 6.95°, a centroid–centroid distance of 3.870 Å, and a shift distance of 1.425 Å (Figure 5d). The Br···Br distances between the Br atom of the C_6H_5Br molecule and the closest bromanilato ligands are 3.82 and 3.90 Å, only slightly above the sum of the van der Waals radii (3.72 Å).

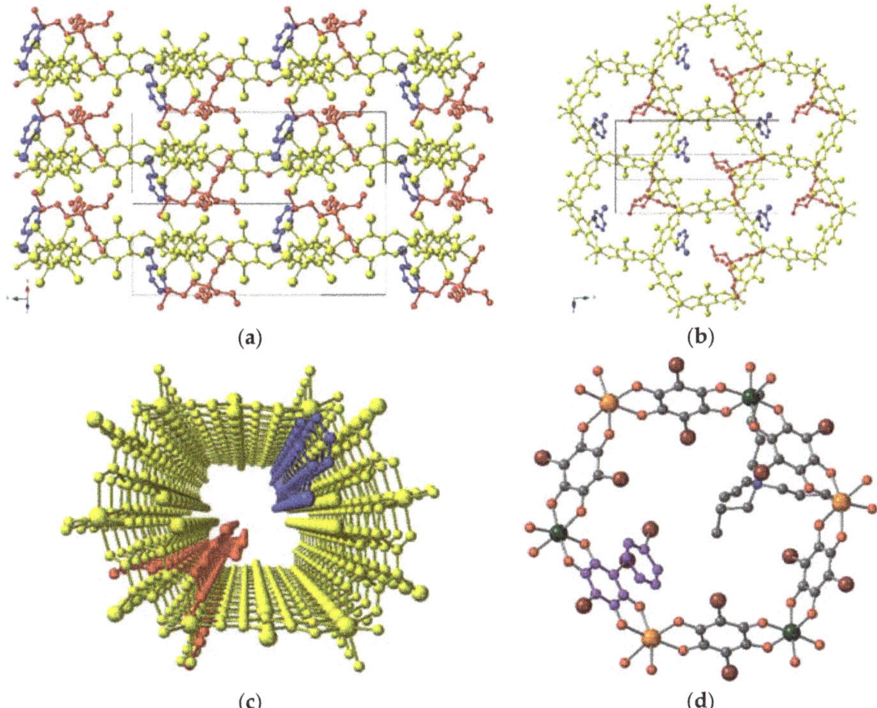

Figure 5. Structure of $(NBu_4)[MnCr(C_6O_4Br_2)_3]·C_6H_5Br·0.5H_2O$ (**A**). (**a**) Side view of the alternating cationic (in red) and anionic (in yellow) layers. The solvent molecules are depicted in blue. (**b**) Top view of the honeycomb layer [same color code as in (**a**)]. (**c**) Perspective view of the hexagonal channels [same color code as in (**a**)]. (**d**) View of one hexagon showing the π–π interactions between the aromatic ring and one anilato ring (in purple). Color code: Mn = orange, Cr = dark green, C = grey, O = red, N = dark blue, and Br = brown. The H atoms were omitted for clarity.

2.5. Magnetic Properties

As expected, compounds **1–6** show quite similar magnetic properties, although, as we note below, there are slight differences in the ordering temperatures and the coercive fields depending on the solvent inserted in the hexagonal channels. All compounds show $\chi_m T$ values at room temperature in the range 6.25–6.35 cm³ K mol⁻¹ (Table 5 and Figure 6a)—very close to the calculated spin only values for S = 3/2 Cr(III) and S = 5/2 high spin Mn(II) ions [χ_m is the molar magnetic susceptibility per Mn(II)Cr(III) couple]. These $\chi_m T$ values show a continuous smooth decrease on cooling the samples and reach minimum values at around 20 K, followed by a sharp increase at around 12 K (inset Figure 6a). This behavior indicates the presence of a ferrimagnetic Mn-Cr coupling in all the samples, as observed in all the previously characterized $[MnCr(C_6O_4X_2)_3]^-$ lattices (Table 1). At *ca.* 10 K, all the samples show a maximum in $\chi_m T$, followed by an abrupt decrease due to the saturation effects of χ_m at low temperatures (Figure 6b). The sharp increase observed at *ca.* 10–12 K suggests the onset of a long range ferrimagnetic ordering, in agreement with the sharp increase observed in the thermal variation of χ_m at *ca.* 10–11 K in all compounds (inset in Figure 6b).

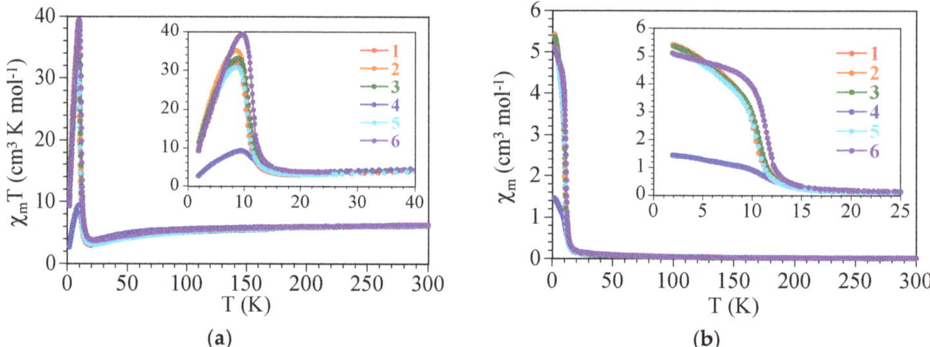

Figure 6. (**a**) Thermal variation of the $\chi_m T$ product (**a**) and χ_m (**b**) for compounds **1–6**. Insets show zooms of the low temperature regions.

The isothermal magnetization cycles at 2 K of all the samples provide a further confirmation of the long range ferrimagnetic ordering (Figure 7). Thus, these measurements show a sharp increase of the magnetization at low fields and hysteresis cycles for all compounds with coercive fields in the range 16.2–56.2 mT (Figure 7b and Table 5). The magnetization values at 5 T in all cases are close to 2.1–2.2 μ_B (Figure 7a and Table 5), which is the expected value for a ferrimagnetic coupling between the S = 3/2 and 5/2 of the Cr(III) and the Mn(II) ions of the lattice. Moreover, at high fields, the magnetization shows a linear smooth increase, further confirming the ferrimagnetic coupling in compounds **1–6**.

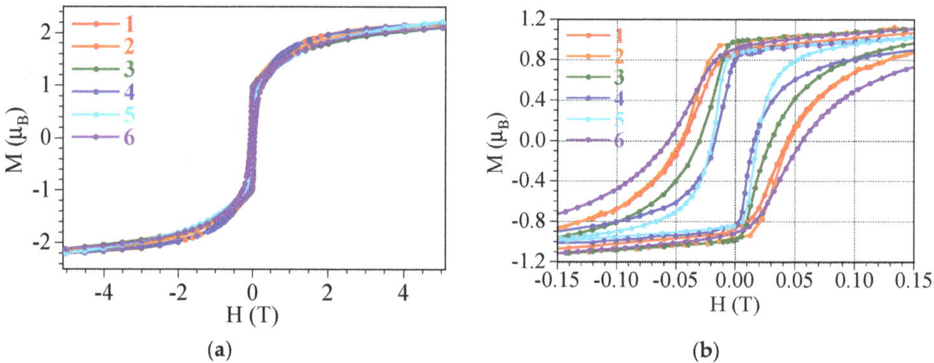

Figure 7. (**a**) Isothermal magnetization cycles at 2 K for compounds **1–6**. (**b**) The low field region.

Table 5. Magnetic properties of compounds **1–6**.

Compound	$\chi_m T$ @ 300 K (cm^3 K mol^{-1})	T_c (K)	M @ 5 T (μ_B)	H_c (mT)
(NBu$_4$)[MnCr(C$_6$O$_4$Cl$_2$)$_3$]·C$_6$H$_5$Cl (**1**)	6.29	10.4	2.20	43.7
(NBu$_4$)[MnCr(C$_6$O$_4$Cl$_2$)$_3$]·1.5C$_6$H$_5$Br (**2**)	6.22	10.7	2.15	44.7
(NBu$_4$)[MnCr(C$_6$O$_4$Cl$_2$)$_3$]·C$_6$H$_5$I (**3**)	6.26	11.0	2.11	30.2
(NBu$_4$)[MnCr(C$_6$O$_4$Cl$_2$)$_3$]· C$_6$H$_5$CH$_3$ (**4**)	6.26	9.8	2.20	16.2
(NBu$_4$)[MnCr(C$_6$O$_4$Cl$_2$)$_3$]·2C$_6$H$_5$CN (**5**)	6.27	10.8	2.20	19.3
(NBu$_4$)[MnCr(C$_6$O$_4$Cl$_2$)$_3$]·2C$_6$H$_5$NO$_2$ (**6**)	6.30	11.2	2.12	56.2

A further confirmation of the long range ferrimagnetic order and a more precise determination of the ordering temperatures was obtained with magnetic susceptibility measurements in the presence of an alternating magnetic field (AC measurements). These measurements show, in all cases, a sharp peak in the in-phase (χ'_m) and in the out-of-phase (χ''_m) signals that does not change with the frequency (Figure 8), confirming the presence of a long range order at low temperatures in all cases. The ordering temperatures, determined as the temperature where χ''_m become non-zero, are all in the range 9.8–11.2 K (Table 5). These values are similar to those observed in most of the reported $[MnCr(C_6O_4X_2)_3]^-$ lattices (Table 1).

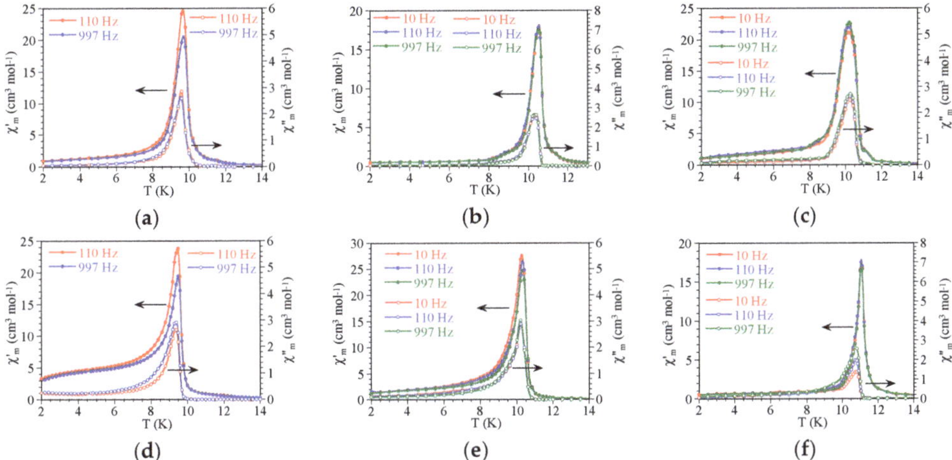

Figure 8. Thermal variation of the in-phase (χ'_m, left scales, filled symbols) and the out-of-phase (χ''_m, right scales, empty symbols) of compounds **1** (**a**), **2** (**b**), **3** (**c**), **4** (**d**), **5** (**e**), and **6** (**f**) at different frequencies.

In order to compare the ordering temperatures (T_c) of compounds **1–6**, we have plotted the thermal variation of χ'_m and χ''_m at a fixed frequency (110 Hz) for all compounds (Figure 9). We can see that, even if the differences in some cases are very small, the order of T_c is: $C_6H_5NO_2$ (**6**) > C_6H_5I (**3**) $\approx$ C_6H_5CN (**5**) > C_6H_5Br (**2**) > C_6H_5Cl (**1**) > $C_6H_5CH_3$ (**4**) (Table 5).

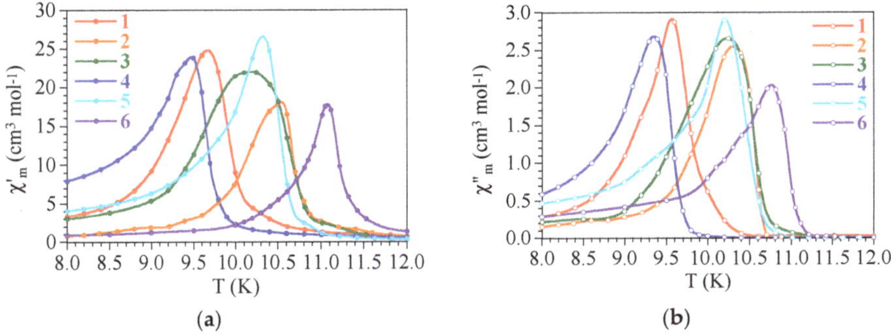

Figure 9. (**a**) Thermal variation of χ'_m (**a**) and χ''_m (**b**) at 110 Hz for compounds **1–6**.

Although we do not have details of the crystal structure, we can assume that, in all cases, the solvent molecules (between one and two per hexagonal cavity) must interact via strong π–π stacking with the anilato rings. This assumption is supported by the similar unit cell parameters determined from the X-ray powder diffractograms (Table 4). If the solvent molecules were located out of the hexagonal cavities (i.e., in the interlayer space), then the *a* and the *c* parameters and the unit cell volume (that are determined by the interlayer space) would be quite different in compounds **1–6**, in contrast with the experimental data. Since T_c depends on the magnetic coupling through the anilato ring, and this coupling depends on the electron density of the anilato rings, [23] we can presume that the differences in T_c reflect the differences in the electron density of the anilato rings (since the six compounds contain the same anilato-type ligand). This modulation of T_c with the electron density on the anilato ring was also observed in the closely related series $(NBu_4)[MnCr(C_6O_4X_2)_3]$ with X = H, Cl, Br and I [23]. On one hand, the sequence observed for the halobenzene derivatives $(C_6H_5I > C_6H_5Br > C_6H_5Cl)$ agrees with the idea that the aromatic ring in C_6H_5Cl had less electron density and, accordingly, donates less electron density to the anilato ring, resulting in a weaker magnetic coupling and a lower T_c. On the other hand, the higher values observed for the C_6H_5CN and $C_6H_5NO_2$ derivatives may be attributed to the fact that there are two aromatic molecules per hexagon in these two compounds. The only compound that do not follow the expected trend is the $C_6H_5CH_3$ derivative (**4**). Since the -CH_3 group is electron donating, it should present a higher T_c than the three halobenzene derivatives. A possible reason to explain this anomaly might be the formation of H-bonds between the -CH_3 group and the oxygen atoms or the chlorine atoms of the chloranilato ligand. The formation of such H-bonds has already been observed in other related Ln(III)-containing anilato-based lattices [34]. The higher temperature needed in the thermogravimetric measurements to release the toluene molecule in this compound agrees with this idea. These H-bonds are expected to shift the aromatic ring of the toluene molecule from its ideal position, reducing the π–π stacking with the anilato ring and, accordingly, the electron density on the anilato ring and T_c.

The idea that the solvent molecules play an important role in T_c is further supported by the fact that the de-solvated compound $(NBu_4)[MnCr(C_6O_4Cl_2)_3]$ (**B**) [23] shows an ordering temperature of 5.5 K (Table 1), well below the observed ones in compounds **1–6**. Although compound **B** has a slightly different structure (the honeycomb layers are alternated, and the hexagonal rings are completely planar), the lower value of T_c in the de-solvated compound suggests that the presence of the solvent molecules increases the electron density in the anilato rings and, accordingly, the magnetic coupling and the ordering temperatures. In fact, preliminary measurements performed on compound **6** after heating the sample at 400 K under vacuum to remove the $PhNO_2$ molecules show a slight decrease in T_c (and an important decrease of the coercive field), further supporting the idea that the solvent molecules are responsible for the fine tuning of T_c.

Despite compounds **1–6** show very close unit cell parameters (Table 4), we cannot discard that, besides the electronic effect, the solvent molecules exert a very tiny structural effect. Although with more important structural changes, this structural effect has already been observed in compounds $(NMe_2H_2)_2[Fe_2(C_6O_4Cl_2)_3]\cdot2H_2O\cdot6DMF$ [27] and $(Et(i\text{-}Pr)_2NH)[MnCr(C_6O_4Br_2)_3]\cdot H_2O\cdot0.5CHCl_3$ [28].

3. Experimental Section

3.1. Starting Materials

Chloranilic acid $(H_2C_6O_4Cl_2)$, $MnCl_2\cdot4H_2O$, and all the used solvents $(C_6H_5Cl, C_6H_5Br, C_6H_5I, C_6H_5CH_3, C_6H_5CN,$ and $C_6H_5NO_2)$ are commercially available and were used as received without further purification. The precursor salt $(NBu_4)_3[Cr(C_6O_4Cl_2)_3]$ was prepared as reported in the literature [23].

3.2. Synthesis of $(NBu_4)[MnCr(C_6O_4Cl_2)_3]\cdot C_6H_5Cl$ (1)

Compound **1** was prepared by carefully layering, at room temperature, a solution of $(NBu_4)_3[Cr(C_6O_4Cl_2)_3]$ (70 mg, 0.05 mmol) in acetonitrile (10 mL) on top of a solution of $MnCl_2\cdot4H_2O$ (40 mg, 0.2 mmol) in 4 mL of MeOH and 6 mL of chlorobenzene. An intermediate layer with a mixture of methanol:chlorobenzene (8:1) was used in order to slow down the diffusion. The solution was allowed to stand for two weeks to obtain a dark powder, which was filtered and air-dried. FT-IR (v/cm^{-1}, KBr pellet): 3440 (vs), 2960 (m), 2930 (w), 2873 (w), 1608 (m), 1497 (vs), 1360 (vs), 1306 (w), 1008 (w), 858 (s), 743 (m), 700 (w), 685 (w), 628 (s), 578 (m), 513 (s), 465 (m).

Anal. Calcd. (%) for $C_{40}H_{41}Cl_7CrMnNO_{12}$: C, 44.37; N, 1.29; H, 3.82. Found (%): C, 43.74; N, 1.21; H, 3.99. Elemental ratio estimated by electron probe microanalysis (EPMA): found: Mn:Cr:Cl = 11.5:11.4:77.1 (1.0:1.0:6.8). Calc. for $C_{40}H_{41}Cl_7CrMnNO_{12}$: Mn:Cr:Cl = 1:1:7.

3.3. Synthesis of (NBu$_4$)[MnCr(C$_6$O$_4$Cl$_2$)$_3$]·1.5C$_6$H$_5$Br (2)

Compound **2** was prepared as **1** but using bromobenzene instead of chlorobenzene. The solution was allowed to stand for two weeks to obtain a dark powder, which was filtered and air-dried. FT-IR (v/cm^{-1}, KBr pellets): 3420 (vs), 2961 (m), 2930 (w), 2872 (w), 1608 (m), 1497 (vs), 1361 (vs), 1305 (w), 1066 (w), 1006 (m), 856 (s), 736 (m), 683 (w), 670 (w), 626 (s), 576 (m), 513 (s), 463 (m).

Anal. Calcd. (%) for $C_{43}H_{43.5}Br_{1.1}Cl_6CrMnNO_{12}$: C, 42.83; N, 1.16; H, 3.63. Found (%): C, 42.82; N, 1.05; H, 3.45. Elemental ratio estimated by electron probe microanalysis (EPMA): found: Mn:Cr:Cl:Br = 9.9:10.2:63.4:16.4 (1.0:1.0:6.4:1.6). Calc. for $C_{43}H_{43.5}Br_{1.5}Cl_6CrMnNO_{12}$: Mn:Cr:Cl:Br = 1:1:6:1.5.

3.4. Synthesis of (NBu$_4$)[MnCr(C$_6$O$_4$Cl$_2$)$_3$]·C$_6$H$_5$I (3)

Compound **3** was prepared as **1** but using iodobenzene instead of chlorobenzene. The solution was allowed to stand for two weeks to obtain a dark powder, which was filtered and air-dried. FT-IR (v/cm^{-1}, KBr pellets): 3443 (vs), 2961 (m), 2926 (w), 2872 (w), 1606 (m), 1496 (vs), 1363 (vs), 1306 (w), 1011 (m), 858 (s), 730 (m), 683 (w), 666 (w), 627 (s), 577 (m), 513 (s), 458 (s).

Anal. Calcd. (%) for $C_{40}H_{41}ICl_6CrMnNO_{12}$: C, 40.91; N, 1.19; H, 3.52. Found (%): C, 39.40; N, 0.99; H, 3.99.

3.5. Synthesis of (NBu$_4$)[MnCr(C$_6$O$_4$Cl$_2$)$_3$]·C$_6$H$_5$CH$_3$ (4)

Compound **4** was prepared as **1** but using toluene instead of chlorobenzene. The solution was allowed to stand for two weeks to obtain a dark powder, which was filtered and air-dried. FT-IR (v/cm^{-1}, KBr pellets): 3440 (vs), 2961 (m), 2930 (w), 2872 (w), 1608 (m), 1497 (vs), 1361 (vs), 1305 (w), 1008 (m), 856 (s), 730 (m), 696 (w), 670 (w), 630 (s), 576 (m), 513 (s), 463 (m).

Anal. Calcd. (%) for $C_{41}H_{44}Cl_6CrMnNO_{12}$: C, 46.35; N, 1.32; H, 4.17. Found (%): C, 45.13; N, 1.12; H, 4.10.

3.6. Synthesis of (NBu$_4$)[MnCr(C$_6$O$_4$Cl$_2$)$_3$]·2C$_6$H$_5$CN (5)

Compound **5** was prepared as **1** but using benzonitrile instead of chlorobenzene. The solution was allowed to stand for two weeks to obtain a dark powder, which was filtered and air-dried. FT-IR (v/cm^{-1}, KBr pellets): 3443 (vs), 2963 (m), 2930 (w), 2873 (w), 1605 (m), 1497 (vs), 1360 (vs), 1306 (w), 1005 (m), 858 (s), 755 (m), 733 (w), 686 (w), 627 (s), 577 (m), 547 (w), 513 (s), 461 (m).

Anal. Calcd. (%) for $C_{44.5}H_{44.5}Cl_6CrMnN_{2.5}O_{12}$: C, 49.00; N, 3.57; H, 3.94. Found (%): C, 47.00; N, 3.16; H, 3.01.

3.7. Synthesis of (NBu$_4$)[MnCr(C$_6$O$_4$Cl$_2$)$_3$]·2C$_6$H$_5$NO$_2$ (6)

Compound **6** was prepared as **1** but using nitrobenzene instead of chlorobenzene. The solution was allowed to stand for two weeks to obtain a dark powder, which was filtered and air-dried. FT-IR (v/cm^{-1}, KBr pellets): 3443 (vs), 2961 (m), 2934 (w), 2872 (w), 1606 (m), 1496 (vs), 1360 (vs), 1306 (w), 1005 (m), 858 (s), 791 (w), 736 (w), 705 (m), 683 (w), 666 (w), 625 (s), 577 (m), 513 (s), 463 (m).

Anal. Calcd. (%) for $C_{43}H_{43.5}Cl_6CrMnN_{2.5}O_{15}$: C, 45.42; N, 3.45; H, 3.81. Found (%): C, 44.56; N, 3.21; H, 3.42.

3.8. Magnetic Measurements

Magnetic measurements were performed with a Quantum Design MPMS-XL-5 SQUID magnetometer (San Diego, CA, USA) with an applied magnetic field of 0.1 T (0.5 T for compound **4**) in the 2–300 K temperature range on polycrystalline samples of all the compounds with masses of 7.514, 3.824, 9.056, 1.002, 5.932, and 4.998 mg for compounds **1–6**, respectively. The isothermal magnetization hysteresis measurements were done with fields from −5 to 5 Tesla at 2 K. AC susceptibility measurements were performed on the same samples with an AC field of 0.395 mT in the temperature range 2–14 K and in the frequency range 10–1000 Hz. Susceptibility data were corrected for the sample holder and for the diamagnetic contribution of the salts using Pascal's constants [46].

3.9. X-ray Powder Diffraction

The X-ray powder diffractograms were collected on polycrystalline samples of compounds **1–6** using a 0.5 mm glass capillary that was mounted and aligned on an Empyrean PANalytical powder diffractometer using CuKα radiation (λ = 1.54177 Å). A total of six scans were collected at room temperature in the 2θ range 5–40° and merged in a single diffractogram.

3.10. Physical Properties

FT-IR spectra were performed on KBr pellets and collected with a Bruker Equinox 55 spectrophotometer. C, H, and N analyses were performed with a Thermo Electron CHNS Flash 2000 analyser and with a Carlo Erba mod. EA1108 CHNS analyzer. The Mn:Cr:Cl:X ratios (X = Cl in **1** and Br in **2**) of the bulk samples were estimated by electron probe microanalysis (EPMA) performed in a Philips SEM XL30 equipped with an EDAX DX-4 microprobe. Thermogravimetric (TG) measurements were performed in Pt crucibles with a TA instruments TGA 550 thermobalance equipped with an autosampler. The TG measurements were done in the 30–700 °C temperature range at 10 °C/min under a N_2 flux of 60 mL/min.

4. Conclusions

The series of six compounds formulated as $(NBu_4)[MnCr(C_6O_4Cl_2)_3]\cdot nG$ with n = 1 for G = C_6H_5Cl (**1**), C_6H_5I (**3**), $C_6H_5CH_3$ (**4**), n = 1.5 for C_6H_5Br (**2**), and n = 2 for G = C_6H_5CN (**5**), and $C_6H_5NO_2$ (**6**) show that it is possible to prepare a complete series of isostructural 2D ferrimagnets by simply changing the solvent molecules. This series presents the classical honeycomb 2D lattice with an eclipsed packing of the layers giving rise to hexagonal channels and shows long range ferrimagnetic order with T_c ranging from 9.8 to 11.2 K. Interestingly, this fine modulation of T_c seems to be related to the electronic character and the number of solvent molecules that enter in the hexagonal channels of these 2D ferrimagnets. We have also prepared the corresponding bromanilato series with the same solvent molecules. Preliminary magnetic measurements show that they are also ferrimagnets with very similar T_c and with a very close sequence of ordering temperatures, further supporting the above results. Work is in progress to complete these series with other aromatic (and not aromatic) solvents and for other anilato ligands and even metal ions. Finally, attempts to remove and/or exchange the different solvent molecules are under investigation. Preliminary results suggest that it is indeed possible to exchange and even to remove the solvents molecules and, as expected, the exchanged and the evacuated compounds slightly modify their ordering temperatures.

Author Contributions: Conceptualization, S.B.; Data curation, C.M.-H., S.B. and C.J.G.-G.; Formal analysis, C.M.-H., S.B. and C.J.G.-G.; Funding acquisition, C.J.G.-G.; Investigation, C.M.-H., S.B. and C.J.G.-G.; Methodology, C.M.-H. and S.B.; Project administration, C.J.G.-G.; Supervision, S.B.; Visualization, S.B.; Writing—original draft, C.J.G.-G.; Writing—review & editing, S.B.

Funding: This research was funded by the Spanish MINECO (project CTQ2017-87201-P AEI/FEDER, UE). C.M.-H. and a pre-doctoral grant from the University of Valencia.

Acknowledgments: Thanks for the funding from the Spanish MINECO (project CTQ2017-87201-P AEI/FEDER, UE). C.M.-H. and the University of Valencia for a pre-doctoral grant.

Conflicts of Interest: The authors declare no conflict of interest.

References

1. Miller, J.S.; Gatteschi, D. Molecule-based magnets. *Chem. Soc. Rev.* **2011**, *40*, 3065–3066. [CrossRef] [PubMed]
2. Verdaguer, M.; Bleuzen, A.; Marvaud, V.; Vaissermann, J.; Seuleiman, M.; Desplanches, C.; Scuiller, A.; Train, C.; Garde, R.; Gelly, G.; et al. Molecules to build solids: High TC molecule-based magnets by design and recent revival of cyano complexes chemistry. *Coord. Chem. Rev.* **1999**, *190*, 1023–1047. [CrossRef]
3. Ohba, M.; Ōkawa, H. Synthesis and magnetism of multi-dimensional cyanide-bridged bimetallic assemblies. *Coord. Chem. Rev.* **2000**, *198*, 313–328. [CrossRef]
4. Culp, J.T.; Park, J.H.; Frye, F.; Huh, Y.D.; Meisel, M.W.; Talham, D.R. Magnetism of metal cyanide networks assembled at interfaces. *Coord. Chem. Rev.* **2005**, *249*, 2642–2648. [CrossRef]
5. Weihe, H.; Güdel, H.U. Magnetic Exchange Across the Cyanide Bridge. *Comments Inorg. Chem.* **2000**, *22*, 75–103. [CrossRef]
6. Ruiz, E.; Rodríguez-Fortea, A.; Alvarez, S.; Verdaguer, M. Is it possible to get high TC magnets with Prussian blue analogues? A theoretical prospect. *Chem. Eur. J.* **2005**, *11*, 2135–2144. [CrossRef] [PubMed]
7. Sato, O.; Iyoda, T.; Fujishima, A.; Hashimoto, K. Photoinduced Magnetization of a Cobalt-Iron Cyanide. *Science* **1996**, *272*, 704–705. [CrossRef] [PubMed]
8. Sato, O. Switchable molecular magnets. *Proc. Jpn. Acad. Ser. B Phys. Biol. Sci.* **2012**, *88*, 213–225. [CrossRef] [PubMed]
9. Shimamoto, N.; Ohkoshi, S.; Sato, O.; Hashimoto, K. Control of Charge-Transfer-Induced Spin Transition Temperature on Cobalt-Iron Prussian Blue Analogues. *Inorg. Chem.* **2002**, *41*, 678–684. [CrossRef]
10. Miyasaka, H.; Julve, M.; Yamashita, M.; Clérac, R. Slow Dynamics of the Magnetization in One-Dimensional Coordination Polymers: Single-Chain Magnets. *Inorg. Chem.* **2009**, *48*, 3420–3437. [CrossRef] [PubMed]
11. Funck, K.E.; Hilfiger, M.G.; Berlinguette, C.P.; Shatruk, M.; Wernsdorfer, W.; Dunbar, K.R. Trigonal-Bipyramidal Metal Cyanide Complexes: A Versatile Platform for the Systematic Assessment of the Magnetic Properties of Prussian Blue Materials. *Inorg. Chem.* **2009**, *48*, 3438–3452. [CrossRef] [PubMed]
12. Ferlay, S.; Mallah, T.; Ouahes, R.; Veillet, P.; Verdaguer, M.; Ouah, R. A room-temperature organometallic magnet based on Prussian blue. *Nature* **1995**, *378*, 701–703. [CrossRef]
13. Tamaki, H.; Zhong, Z.J.; Matsumoto, N.; Kida, S.; Koikawa, M.; Achiwa, N.; Hashimoto, Y.; Okawa, H. Design of metal-complex magnets. Syntheses and magnetic properties of mixed-metal assemblies {NBu$_4$[MCr(ox)$_3$]}$_x$ (NBu$_4{}^+$ = tetra(n-butyl)ammonium ion; ox^{2-} = oxalate ion; M = Mn^{2+}, Fe^{2+}, Co^{2+}, Ni^{2+}, Cu^{2+}, Zn^{2+}). *J. Am. Chem. Soc.* **1992**, *114*, 6974–6979. [CrossRef]
14. Mathoniere, C.; Nuttall, C.J.; Carling, S.G.; Day, P. Ferrimagnetic Mixed-Valency and Mixed-Metal Tris(oxalato)iron(III) Compounds: Synthesis, Structure, and Magnetism. *Inorg. Chem.* **1996**, *35*, 1201–1206. [CrossRef] [PubMed]
15. Clement, R.; Decurtins, S.; Gruselle, M.; Train, C. Polyfunctional two-(2D) and three-(3D) dimensional oxalate bridged bimetallic magnets. *Monatshefte Chem.* **2003**, *134*, 117–135. [CrossRef]
16. Benmansour, S.; Cerezo-Navarrete, C.; Canet-Ferrer, J.; Muñoz-Matutano, G.; Martínez-Pastor, J.; Gómez-García, C.J. A fluorescent layered oxalato-based canted antiferromagnet. *Dalton Trans.* **2018**, *47*, 11909–11916. [CrossRef] [PubMed]
17. Clemente-León, M.; Coronado, E.; Galán-Mascarós, J.R.; Gómez-Garcia, C.J. Intercalation of decamethylferrocenium cations in bimetallic oxalate-bridged two-dimensional magnets. *Chem. Commun.* **1997**, 1727–1728.
18. Coronado, E.; Clemente-León, M.; Galán-Mascarós, J.R.; Giménez-Saiz, C.; Gómez-García, C.J.; Martínez-Ferrero, E. Design of molecular materials combining magnetic, electrical and optical properties. *J. Chem. Soc. Dalton Trans.* **2000**, 3955–3961. [CrossRef]
19. Coronado, E.; Galán-Mascarós, J.R.; Gómez-García, C.J.; Ensling, J.; Gütlich, P. Hybrid molecular magnets obtained by insertion of decamethyl-metallocenium cations into layered, bimetallic oxalate complexes: [Z^{III}Cp*$_2$][M^{II}M^{III}(ox)$_3$](Z^{III} = Co, Fe; M^{III} = Cr, Fe; M^{II} = Mn, Fe, Co, Cu, Zn; ox = oxalate; Cp* = pentamethylcyclopentadienyl). *Chem. Eur. J.* **2000**, *6*, 552–563. [CrossRef]

20. Coronado, E.; Galán-Mascarós, J.R.; Gómez-García, C.J.; Martínez-Agudo, J.M. Increasing the coercivity in layered molecular-based magnets A[$M^{II}M^{III}$(ox)$_3$] (M^{II} = Mn, Fe, Co, Ni, Cu; M^{III} = Cr, Fe; ox = oxalate; A = organic or organometallic cation). *Adv. Mater.* **1999**, *11*, 558–561. [CrossRef]

21. Coronado, E.; Galán-Mascarós, J.R.; Gómez-García, C.J.; Martínez-Agudo, J.M.; Martínez-Ferrero, E.; Waerenborgh, J.C.; Almeida, M. Layered molecule-based magnets formed by decamethylmetallocenium cations and two-dimensional bimetallic complexes [$M^{II}Ru^{III}$(ox)$_3$]$^-$ (M^{II} = Mn, Fe, Co, Cu and Zn; ox = oxalate). *J. Solid State Chem.* **2001**, *159*, 391–402. [CrossRef]

22. Coronado, E.; Galán-Mascarós, J.R.; Gómez-García, C.J.; Laukhin, V. Coexistence of ferromagnetism and metallic conductivity in a molecule-based layered compound. *Nature* **2000**, *408*, 447–449. [CrossRef] [PubMed]

23. Atzori, M.; Benmansour, S.; Mínguez Espallargas, G.; Clemente-León, M.; Abhervé, A.; Gómez-Claramunt, P.; Coronado, E.; Artizzu, F.; Sessini, E.; Deplano, P.; et al. A Family of Layered Chiral Porous Magnets Exhibiting Tunable Ordering Temperatures. *Inorg. Chem.* **2013**, *52*, 10031–10040. [CrossRef] [PubMed]

24. Mercuri, M.L.; Congiu, F.; Concas, G.; Sahadevan, S.A. Recent Advances on Anilato-Based Molecular Materials with Magnetic and/or Conducting Properties. *Magnetochem* **2017**, *3*, 17. [CrossRef]

25. Kitagawa, S.; Kawata, S. Coordination compounds of 1,4-dihydroxybenzoquinone and its homologues. Structures and properties. *Coord. Chem. Rev.* **2002**, *224*, 11–34. [CrossRef]

26. Halis, S.; Inge, A.K.; Dehning, N.; Weyrich, T.; Reinsch, H.; Stock, N. Dihydroxybenzoquinone as Linker for the Synthesis of Permanently Porous Aluminum Metal–Organic Frameworks. *Inorg. Chem.* **2016**, *55*, 7425–7431. [CrossRef]

27. Jeon, I.R.; Negru, B.; Van Duyne, R.P.; Harris, T.D. A 2D Semiquinone Radical-Containing Microporous Magnet with Solvent-Induced Switching from Tc = 26 to 80 K. *J. Am. Chem. Soc.* **2015**, *137*, 15699–15702. [CrossRef]

28. Palacios-Corella, M.; Fernández-Espejo, A.; Bazaga-García, M.; Losilla, E.R.; Cabeza, A.; Clemente-León, M.; Coronado, E. Influence of Proton Conducting Cations on the Structure and Properties of 2D Anilate-Based Magnets. *Inorg. Chem.* **2017**, *56*, 13865–13877. [CrossRef]

29. Riley, P.E.; Haddad, S.F.; Raymond, K.N. Preparation of praseodymium(III) chloranilate and the crystal structures of Pr$_2$(C$_6$Cl$_2$O$_4$)$_3$.8C$_2$H$_5$OH and Na$_3$[C$_6$H$_2$O(OH)(SO$_3$)$_2$].H$_2$O. *Inorg. Chem.* **1983**, *22*, 3090–3096. [CrossRef]

30. Christian, R. Complexes with substituted 2,5-dihydroxy-p-benzoquinones: The inclusion compounds [Y(H$_2$O)$_3$]$_2$(C$_6$Cl$_2$O$_4$)$_3$·6.6H$_2$O and [Y(H$_2$O)$_3$]$_2$(C$_6$Br$_2$O$_4$)$_3$·6H$_2$O. *Mater. Res. Bull.* **1987**, *22*, 1483–1491.

31. Abrahams, B.F.; Coleiro, J.; Hoskins, B.F.; Robson, R. Gas hydrate-like pentagonal dodecahedral M$_2$(H$_2$O)$_{18}$ cages (M = lanthanide or Y) in 2,5-dihydroxybenzoquinone-derived coordination polymers. *Chem. Commun.* **1996**, 603–604. [CrossRef]

32. Coleiro, J.; Ha, K.; Orchard, S.D.; Robson, R.; Abrahams, B.F.; Hoskins, B.F. Dihydroxybenzoquinone and chloranilic acid derivatives of rare earth metals. *J. Chem. Soc. Dalton Trans.* **2002**, 1586–1594.

33. Benmansour, S.; Pérez-Herráez, I.; López-Martínez, G.; Gómez-García, C.J. Solvent-modulated structures in anilato-based 2D coordination polymers. *Polyhedron* **2017**, *135*, 17–25. [CrossRef]

34. Benmansour, S.; Pérez-Herráez, I.; Cerezo-Navarrete, C.; López-Martínez, G.; Martínez Hernández, C.; Gómez-García, C.J. Solvent-modulation of the structure and dimensionality in lanthanoid-anilato coordination polymers. *Dalton Trans.* **2018**, *47*, 6729–6741. [CrossRef] [PubMed]

35. Benmansour, S.; Hernández-Paredes, A.; Gómez-García, C.J. Effect of the lanthanoid-size on the structure of a series of lanthanoid-anilato 2-D lattices. *J. Coord. Chem.* **2018**, *71*, 845–863. [CrossRef]

36. Gómez-Claramunt, P.; Benmansour, S.; Hernández-Paredes, A.; Cerezo-Navarrete, C.; Rodríguez-Fernández, C.; Canet-Ferrer, J.; Cantarero, A.; Gómez-García, C.J. Tuning the Structure and Properties of Lanthanoid Coordination Polymers with an Asymmetric Anilato Ligand. *Magnetochem* **2018**, *4*, 6.

37. Benmansour, S.; Hernández-Paredes, A.; Gómez-García, C.J. Two Dimensional Magnetic Coordination Polymers Formed by Lanthanoids and Chlorocyananilato. *Magnetochem* **2018**, *4*, 58. [CrossRef]

38. Ashoka Sahadevan, S.; Monni, N.; Abhervé, A.; Marongiu, D.; Sarritzu, V.; Sestu, N.; Saba, M.; Mura, A.; Bongiovanni, G.; Cannas, C.; et al. Nanosheets of Two-Dimensional Neutral Coordination Polymers Based on Near-Infrared-Emitting Lanthanides and a Chlorocyananilate Ligand. *Chem. Mater.* **2018**, *30*, 6575–6586. [CrossRef]

39. Gómez-Claramunt, P. Anilato-Based Multifunctional Molecular Materials. Ph.D. Thesis, University of Valencia, Valencia, Spain, 2018.

40. Abhervé, A.; Verneret, M.; Clemente-León, M.; Coronado, E.; Gómez-García, C.J. One-Dimensional and Two-Dimensional Anilate-Based Magnets with Inserted Spin-Crossover Complexes. *Inorg. Chem.* **2014**, *53*, 12014–12026.

41. Abhervé, A.; Mañas-Valero, S.; Clemente-León, M.; Coronado, E. Graphene related magnetic materials: Micromechanical exfoliation of 2D layered magnets based on bimetallic anilate complexes with inserted $[Fe^{III}(acac_2\text{-trien})]^+$ and $[Fe^{III}(sal_2\text{-trien})]^+$ molecules. *Chem. Sci.* **2015**, *6*, 4665–4673. [CrossRef]

42. Benmansour, S.; Abhervé, A.; Gómez-Claramunt, P.; Vallés-García, C.; Gómez-García, C.J. Nanosheets of Two-Dimensional Magnetic and Conducting Fe(II)/Fe(III) Mixed-Valence Metal–Organic Frameworks. *ACS Appl. Mater. Interfaces* **2017**, *9*, 26210–26218. [CrossRef] [PubMed]

43. Sahadevan, S.A.; Abhervé, A.; Monni, N.; Sáenz, D.P.; Galán-Mascarós, J.R.; Waerenborgh, J.C.; Vieira, B.J.C.; Auban-Senzier, P.; Pillet, S.; Bendeif, E.; et al. Conducting Anilate-Based Mixed-Valence Fe(II)Fe(III) Coordination Polymer: Small-Polaron Hopping Model for Oxalate-Type Fe(II)Fe(III) 2D Networks. *J. Am. Chem. Soc.* **2018**, *140*, 12611–12621. [CrossRef] [PubMed]

44. Martínez-Hernández, C.; Benmansour, S.; Gómez-García, C.J. Modulation of the ordering temperature in anilato-based magnets. *Polyhedron* **2019**, in press. [CrossRef]

45. *PANalytical X'Pert HighScore Plus*; PANalytical l. V.: Almelo, The Netherlands, 2012.

46. Bain, G.A.; Berry, J.F. Diamagnetic Corrections and Pascal's Constants. *J. Chem. Educ.* **2008**, *85*, 532. [CrossRef]

 *magnetochemistry*

Article

A Novel Family of Triangular $Co^{II}_2Ln^{III}$ and $Co^{II}_2Y^{III}$ Clusters by the Employment of Di-2-Pyridyl Ketone [†]

Constantinos G. Efthymiou [1], Áine Ní Fhuaráin [1], Júlia Mayans [2], Anastasios Tasiopoulos [3], Spyros P. Perlepes [4] and Constantina Papatriantafyllopoulou [1,*]

[1] SSPC, Synthesis and Solid State Pharmaceutical Centre, School of Chemistry, National University of Ireland Galway, University Road, H91 TK33 Galway, Ireland; dinosef@yahoo.com (C.G.E.); A.NIFHUARAIN1@nuigalway.ie (Á.N.F.)

[2] Departament de Quimica Inorgànica i Orgànica, Secció Inorgànica and Institut de Nanociencia I, Nanotecnologia (IN2UB), Universitat de Barcelona, Marti i Franquès 1-11, 08028 Barcelona, Spain; julia.mayans@qi.ub.edu

[3] Department of Chemistry, University of Cyprus, 1678 Nicosia, Cyprus; atasio@ucy.ac.cy

[4] Department of Chemistry, University of Patras, 26504 Patras, Greece; perlepes@patreas.upatras.gr

* Correspondence: constantina.papatriantafyllopo@nuigalway.ie; Tel.: +353-91-493462

† This paper is dedicated to Professor Masahiro Yamashita, a leading and inspiring Scientist on Molecular Magnetism, on the occasion of his 65th birthday.

Received: 31 March 2019; Accepted: 24 May 2019; Published: 4 June 2019

Abstract: The synthesis, structural characterization and magnetic study of novel Co^{II}/4f and Co^{II}/Y^{III} clusters are described. In particular, the initial employment of di-2-pyridyl ketone, $(py)_2CO$, in mixed metal Co/4f chemistry, provided access to four triangular clusters, $[Co^{II}_2M^{III}\{(py)_2C(OEt)(O)\}_4(NO_3)(H_2O)]_2[M(NO_3)_5](ClO_4)_2$ (M = Gd, **1**; Dy, **2**; Tb, **3**; Y, **4**), where $(py)_2C(OEt)(O)^-$ is the monoanion of the hemiketal form of $(py)_2CO$. Clusters **1–4** are the first reported Co/4f (**1–3**) and Co/Y (**4**) species bearing $(py)_2CO$ or its derivatives, despite the fact that over 200 metal clusters bearing this ligand have been reported so far. Variable-temperature, solid-state dc and ac magnetic susceptibility studies were carried out on **1–4** and revealed the presence of weak ferromagnetic exchange interactions between the metal ions (J_{Co-Co} = +1.3 and +0.40 cm^{-1} in **1** and **4**, respectively; J_{Co-Gd} = +0.09 cm^{-1} in **1**). The ac susceptibility studies on **2** revealed nonzero, weak out-of-phase (χ''_M) signals below ~5 K.

Keywords: 3d/4f metal clusters; di-2 pyridyl ketone; magnetism; cobalt; lanthanides; mixed metal Co/Ln clusters

---

1. Introduction

The synthesis and characterization of new mixed-metal 3d/4f clusters has attracted immense interest over the last few decades, due to their fascinating structural features (high nuclearities, unprecedented metal topologies, aesthetically pleasing architectures, etc.), as well as due to their interesting magnetic properties [1–3]. In particular, 4f ions often favor the formation of heterometallic compounds that possess exceptionally high nuclearities, with representative examples being clusters of $Ni_{64}Gd_{96}$ [4], $Ni_{76}La_{60}$ [5], $Ni_{54}Gd_{54}$ [6], $Cu_{36}Dy_{24}$ [7], $Ni_{10}Gd_{42}$ [8], $Ni_{30}La_{20}$ [9,10], etc. This intriguing ability of 4f ions possibly stems from their strong oxophilicity, which, in combination with their high coordination numbers, results in the formation of hydroxo/oxo species that readily promote the aggregation process. Concerning the magnetic properties of the 3d/4f compounds, the 4f ions bring several advantages, such as their considerable number of unpaired electrons (e.g., Gd^{3+} has seven unpaired e−) and their large single ion anisotropy (e.g., Tb^{3+}, Dy^{3+}, Ho^{3+}, etc.) as a result of their orbital angular momentum. The above properties make them ideal candidates for the synthesis of heterometallic clusters with single-molecule magnetism behavior (SMMs) [11,12], fulfilling the desirable features for a compound to behave as an SMM, namely (i) high spin ground state (S) and (ii) negative axial zero field splitting parameter (D). SMMs are discrete metal compounds that exhibit superparamagnetic behavior below a blocking temperature T_B and have been proposed for several technological applications including high-density information storage, molecular spintronics and qubits for quantum computation [11–15].

The study of mixed-metal 3d/4f reaction systems, as a means for the isolation of new SMMs with a high energy barrier for the magnetization reversal, has led to a large variety of such species that now include Mn/4f [16–21], Fe/4f [22–25], Ni/4f [8,26], Cu/4f [27–30] and Co/4f [8,31–33] compounds [1,2]. It is noteworthy that the majority of 3d/4f SMMs are Mn/4f clusters containing some Mn^{III} centers with an $S = 2$ spin state and a significant uniaxial anisotropy. Some remarkable examples, e.g., a Mn_6Tb_2 [18] and a $Mn_{21}Dy$ [17] cluster, display high energy barriers for the magnetization reversal ($U_{eff} = 103$ K and 74 K, respectively), which are of comparable magnitude to the family of the most thoroughly studied homometallic carboxylato Mn_{12} SMMs [11]. On the other hand, the reported Co/4f SMMs are significantly less, despite the fact that the combination of the anisotropic $3d^7$ Co^{II} with the 4f ions has a great potential to yield SMMs with high U_{eff} and distinctively different properties from other heterometallic species. A possible explanation for this could be related to synthetic challenges such as the oxidation of the Co^{2+} to the diamagnetic and low spin Co^{3+}, which occurs easily in the presence of a base under ambient conditions.

Many carboxylate and O or N,O-ligands have been used for the synthesis of 3d/4f metal clusters [1–3]; amongst them, di-2-pyridyl ketone ((py)$_2$CO, Scheme 1) is very attractive as its carbonyl group can easily undergo nucleophilic attack, providing a wide range of hemiaketal and *gem*-diol derivatives that are able to link many metal ions, favoring the formation of high nuclearity metal clusters with interesting structural features and magnetic properties [34,35]. Over 200 homo- and heterometallic compounds have now been reported, containing (py)$_2$CO and its derivatives, thus the absence of such Co/4f clusters is noticeable considering the great development of this research field.

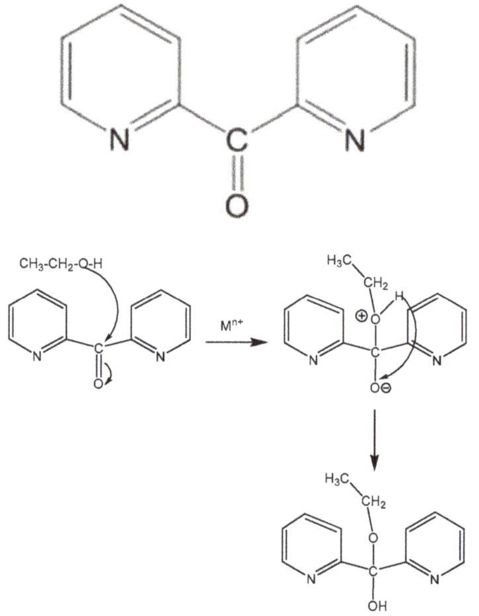

Scheme 1. A schematic representation of (py)$_2$CO (**top**) and its transformation to (py)$_2$C(OEt)(OH) (**bottom**).

In 2010, our groups employed $(py)_2CO$ in Ni/4f cluster chemistry and reported a new family of triangular Ni_2Ln compounds with interesting magnetic properties [36,37]. Expanding this research, we herein report the synthesis and characterization of four isostructural triangular Co_2M clusters (M=Tb, Dy, Gd, Y); these compounds are the first examples of Co/4f or Y species bearing $(py)_2CO$ and its derivatives, with the Co_2Dy analogue displaying out-of-phase ac magnetic susceptibility signals at low temperatures, indicative of the slow relaxation of the magnetization.

2. Results and Discussion

2.1. Synthetic Comments

We have developed an intense interest in the synthesis of 3d/4f metal clusters by the employment of various pyridyl oximate- and alkoxide-containing ligands; these research efforts have yielded a variety of new mixed-metal species with interesting structural features and magnetic properties, including Ni_8Dy_8 [38,39], Ni_2Ln_2 [40], Ni_3Ln [26], Ni_2Ln [36,37,40], Mn_4Ln_2 [41], etc. Restricting further discussion to the use of $(py)_2CO$ in this field, we recently reported the first Mn/4f compounds, which belong to a family of cross-shaped Mn_4Ln_2 clusters, where some of them exhibit slow relaxation of magnetization; whereas, in the past, we reported the first Ni/4f compounds with the monoanionic form of $(py)_2CO$. Wishing to expand this work, we recently decided to investigate the previously unexplored reaction system of $Co^{2+}/Ln^{3+}/(py)_2CO$.

The reaction of $Co(ClO_4)_2 \cdot 6H_2O$, $Ln(NO_3)_3 \cdot 6H_2O$ (Ln=Gd, **1**; Dy, **2**; Tb, **3**) or $Y(NO_3)_3 \cdot 6H_2O$ (**4**), $(py)_2CO$ and $CH_3CO_2Na \cdot 3H_2O$ in EtOH afforded a red solution from which well-shaped red crystals of compounds **1–4** with the general formula $[Co_2M\{(py)_2C(OEt)(O)\}_4(NO_3)(H_2O)]_2[M(NO_3)_5](ClO_4)_2$ were subsequently isolated. The formation of **1–4** is summarized in Equation (1).

$$4\,Co(ClO_4)_2 \cdot 6H_2O + 3M(NO_3)_3 \cdot 6H_2O + 8(py)_2CO + 8NaO_2CMe \cdot 3H_2O + 8EtOH \xrightarrow{EtOH}$$
$$\left[Co_2M\{(py)_2C(OEt)(O)\}_4(NO_3)(H_2O)\right]_2[M(NO_3)_5](ClO_4)_2 + 8MeCO_2H + 2NaNO_3 + 6NaClO_4 + 66H_2O \quad (1)$$
$$Gd, 1; Dy, 2; Tb, 3; Y, 4$$

The nature of the base and the crystallization method are not crucial for the identity of the products and affect only their crystallinity and the reaction yield; we were able to isolate **1–4** (IR evidence) by using other bases, such as NaOMe, NaOEt, $LiOH \cdot H_2O$, etc. On the other hand, the ratio of the reactants and the nature of solvent affect the product identity, as by further increasing the excess of $(py)_2CO$, mononuclear Co^{II} compounds are isolated. EtOH is the only solvent that favors the formation of **1–4**, whereas the use of different solvents yields amorphous products that could not be further characterized.

2.2. Description of Structures

A representation of the cationic $[Co_2Gd\{(py)_2C(OEt)(O)\}_4(NO_3)(H_2O)]^{2+}$ that is present in the molecular structure of **1** is shown in Figure 1. A representation of the elipsoid plot for **1** is shown in Figure S1 in the supplementary material. Selected interatomic distances and angles for **1** are listed in Table 1.

Complex **1** crystallizes in the monoclinic space group $C\,2/c$. Its structure consists of two isostructural triangular cationic clusters $[Co_2Gd\{(py)_2C(OEt)(O)\}_4(NO_3)(H_2O)]^{2+}$, which are symmetrically related with a 2-fold crystallographic axis. The positive charge of the cation is balanced by one $[Gd(NO_3)_5]^{2-}$ and two NO_3^- counterions. The cationic cluster is comprised of two Co^{2+} and one Gd^{3+} ions, which are held together by four $(py)_2C(OEt)(O)^-$ ligands. The $\{Co_2GdO_4\}^{3+}$ core of this complex displays an oxo-centered triangular arrangement, in which one μ_3-alkoxo group coming from one $(py)_2C(OEt)(O)^-$ ligand bridges the three metal centres; in addition, three μ_2-O^{2-} ions, from three different $(py)_2C(OEt)(O)^-$ ligands, are located peripherally, bridging the two metal ions in each edge of the triangle. Alternatively, the structural core in **1** can be described as a defective cubane, in which one vertex and three edges are missing. The central μ_3-O^{2-} ion deviates 1.12(2) Å from the plane formed by the metal ions. The intermetallic distances in **1** are Co1 … Gd = 3.471 Å, Co2 … Gd = 3.546 Å and Co1 … Co2 = 3.192 Å.

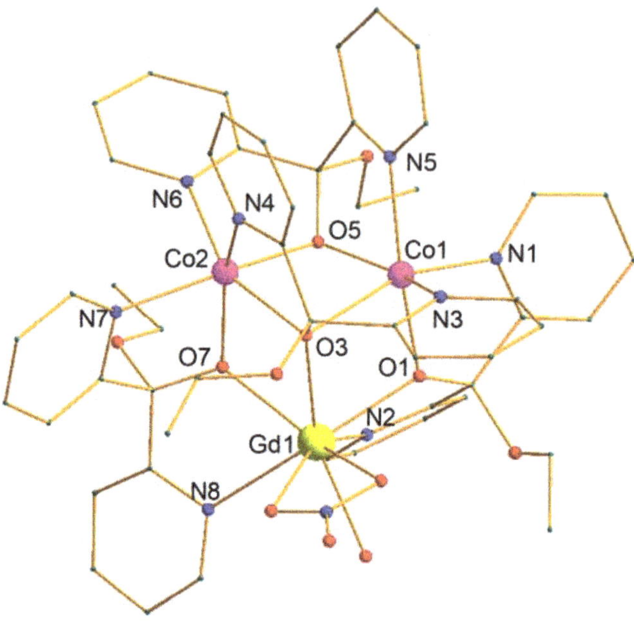

Figure 1. Structure of the cationic cluster **1**. The hydrogen atoms and the counteranions were omitted for clarity.

Table 1. Selected interatomic distances (Å) and angles (degrees) for **1**.

Gd(1)-O(1)	2.252(15)	Co(1)-N(5)	2.220(20)
Gd(1)-O(3)	2.387(15)	Co(2)-O(3)	2.142(13)
Gd(1)-O(7)	2.304(13)	Co(2)-O(5)	2.048(14)
Gd(1)-N(2)	2.582(17)	Co(2)-O(7)	2.046(14)
Gd(1)-N(8)	2.578(19)	Co(2)-N(4)	2.070(16)
Co(1)-O(1)	2.021(15)	Co(2)-N(6)	2.08(2)
Co(1)-O(3)	2.265(15)	Co(2)-N(7)	2.176(19)
Co(1)-O(5)	1.984(16)	Gd(1)-Co(1)	3.471(3)
Co(1)-N(1)	2.083(18)	Gd(1)-Co(2)	3.546(3)
Co(1)-N(3)	2.093(19)	Co(1)-Co(2)	3.192(4)
Co(1)-O(1)-Gd(1)	108.5(6)	Co(2)-O(7)-Gd(1)	109.1(5)
Co(1)-O(3)-Gd(1)	96.5(5)	Co(1)-O(3)-Co(2)	92.8(5)
Co(2)-O(3)-Gd(1)	102.9(6)	Co(1)-O(5)-Co(2)	104.6(7)

The monoanionic $(py)_2C(OEt)(O)^-$ ligands are derived from the nucleophilic attack of one EtOH molecule on the central C atom of the carbonyl group of $(py_2)CO$. The three $(py)_2C(OEt)(O)^-$ ligands adopt a $\eta^1{:}\eta^2{:}\eta^1{:}\mu_2$ coordination mode, with the fourth one being coordinated to the metals in a $\eta^1{:}\eta^3{:}\eta^1{:}\mu_3$ fashion (Scheme 2). The two Co^{II} ions are six-coordinate with their coordination spheres ({O1, O5, O3, N1, N3, N5} for Co1; {O3, O5, O7, N4, N6, N7} for Co2) displaying distorted octahedral geometries. The three O and the three N donor atoms around each Co^{II} ion adopt the facial, fac-topological arrangement; each Co^{II} ion is surrounded by three five-membered chelate rings, formed by three different $(py)_2C(OEt)(O)^-$ ligands. The Co oxidation state was assigned by charge considerations and bond-valence sum (BVS) calculations [42].

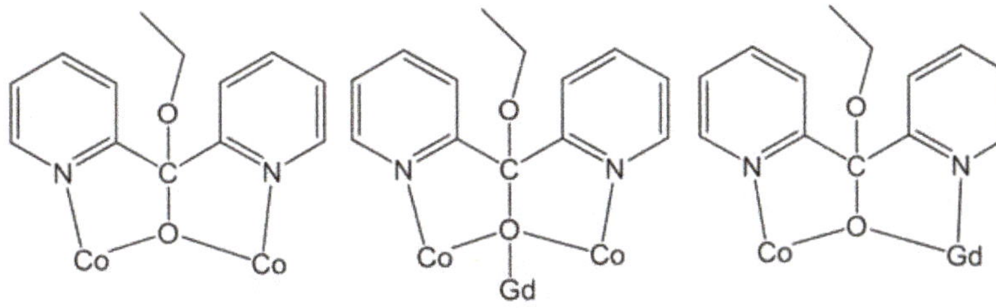

$$\eta^1{:}\eta^2{:}\eta^1{:}\mu_2 \qquad\qquad \eta^1{:}\eta^3{:}\eta^1{:}\mu_3 \qquad\qquad \eta^1{:}\eta^2{:}\eta^1{:}\mu_2$$

Scheme 2. A schematic representation of the coordination modes of $(py)_2C(OEt)(O)^-$ in **1**.

Gd1 is eight-coordinate and its {O1, O3, O7, O9, O10, O12, N2, N7} coordination sphere is rich in O donor atoms as a consequence of the oxophilic character of the lanthanides. Its coordination environment is formed by two five-membered chelate rings, the central μ_3-O^{2-} ion, one bidentate chelate NO_3^- ion and one terminal H_2O molecule. Gd2 in the $[Gd(NO_3)_5]^{2-}$ ion is 10-coordinated, surrounded by five bidentate chelating nitrate groups. Gd2 lies on a crystallographic 2-fold axis, which passes through the N11 atom of a NO_3^- group.

To deduce the coordination polyhedra defined by the donor atoms around Gd1, a comparison of the experimental structural data with the theoretical data for the most common polyhedral structures with eight vertices was performed by means of the program SHAPE [43,44]; a reliable, high-quality fit was not achieved.

Closer inspection of the crystal structure of **1** reveals the absence of strong H-bonding interactions. This might be a result of the very well-separated neighboring Co_2Gd units. The shortest metal$\cdots$metal distance between neighboring trinuclear clusters is 10.564 Å (Gd1 ... Gd1), while the shortest metal$\cdots$metal distance between a trinuclear cluster with a neighboring $[Gd(NO_3)_5]^{2-}$ anion is 7.491 Å (Gd1 ... Gd2).

Compounds **2–4** are isostructural with **1**, as confirmed by a comparison of their unit cell dimensions. The identity, purity and stability of these compounds was also studied by powder X-ray diffraction (pxrd) studies (Figure S2 in the Supplementary Material).

Compound **1** and its structural analogues (**2–4**) are the first Co/Ln or Y clusters bearing $(py)_2CO$ and/or its transformed gem-diol or hemiketal derivatives. They also join the very small family of heterometallic 3d/4f/$(py)_2CO$ clusters [26,36,37,41,45,46]; thus, they provide insight into the coordination chemistry of this versatile ligand and unlock the chemical and structural features, which can further lead to the isolation of higher nuclearity heterometallic species.

2.3. Magnetism Studies

Solid-state, variable-temperature dc magnetic susceptibility (χ_M) data were collected on vacuum-dried microcrystalline samples of complexes **1–4** in the 2.0–300 K range, and they are shown in Figure 2, top, as $\chi_M T$ vs. T plots. The experimental values for **1–4** at 300 K are 16.04, 26.86, 23.09 and 5.5 $cm^3 \cdot K \cdot mol^{-1}$, respectively, being close to the expected ones for one and a half non-interacting Ln^{III} cations (**1**, Gd, S = 7/2, L = 0, $^8S_{7/2}$, g = 2; **2**, Dy, S = 5/2, L = 5, $^6H_{15/2}$, g = 4/3; **3**, Tb, S = 3, L = 3, $^7F_6$, g = 3/2; **4**, Y, S = 0) and two non-interacting high spin Co^{II} cations (S = 3/2, g = 2) of 15.05, 26.1, 21.5 and 3.8 $cm^3 \cdot K \cdot mol^{-1}$, respectively.

The study of the static magnetic properties of highly anisotropic Ln^{III} cations with high-spin Co^{II} ions (S = 3/2) within the same molecule is challenging because both types of paramagnetic centers present spin-orbit contribution due to the strong orbital contribution to the magnetic moment; this yields high anisotropies, which prevent the use of spin-only Hamiltonians for the mathematical interpretation and fitting of the experimental curves [47,48]. Although L is not fully quenched, spin-only Hamiltonians are used to fit the curves for practical reasons, where feasible, in the reported compounds.

For complex **4**, the $\chi_M T$ vs. T curve remains almost constant with the decreasing temperature from 300 to 50 K and then drops to 3.7 cm^3·K·mol^{-1}. This complex contains a diamagnetic Y^{III}, which allows the study of the interaction between the Co(II) ions using the spin Hamiltonian H = $-2J$ ($\hat{S}_{Co1}·\hat{S}_{Co2}$) + $D\hat{S}_{z2}$ + $\Sigma_i \mu g_{eff}\vec{H}\hat{S}_i$ in the full range of temperature; the exchange interactions between the Co^{II} ions are weak ferromagnetic with J = +0.40 cm^{-1}, D = 9.5 cm^{-1} and g = 2.35.

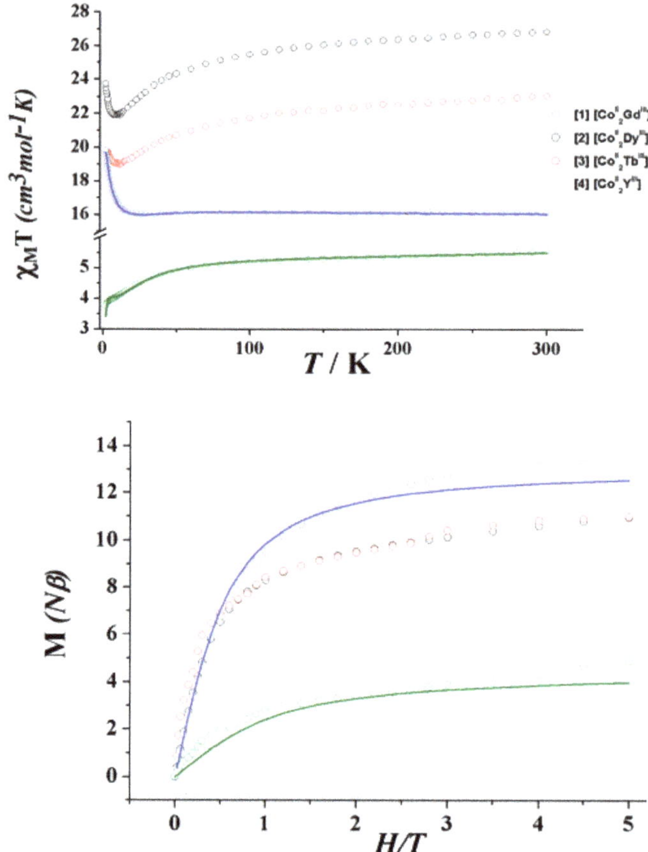

Figure 2. $\chi_M T$ vs. T plots (**top**) and field dependence of the magnetization at 2 K (**bottom**) for **1–4**. Solid line represents the best fit for **1** and **4**.

The $\chi_M T$ vs. T curve for **1** remains almost constant until 20 K and then starts to increase, reaching the value of 20.01 cm^3·K·mol^{-1} at 2 K, which shows an extremely weak ferromagnetic coupling between the metal ions. The fitting of the experimental data to the Hamiltonian equation H = $-2J$($\hat{S}_{Co1}$ $\hat{S}_{Gd}$ + $\hat{S}_{Co2}$ $\hat{S}_{Gd}$)$-2J'$($\hat{S}_{Co1}$ $\hat{S}_{Co2}$) + $D\hat{S}_z^2$ + $\Sigma_i \mu g_{eff}\vec{H}\hat{S}_i$, in the whole temperature range, provided the coupling values between Co^{II}-Co^{II} ions (J = +1.3 cm^{-1}) and Co^{II}-Gd^{III} ions (J = +0.09 cm^{-1}), respectively, with a mean g value of 2.35. This magnetic coupling is in agreement with previous studies in Co^{II}-Gd^{III} complexes, which always present a ferromagnetic coupling when the Co^{II} is a high spin cation [49,50].

Complexes **2** and **3** exhibit a similar magnetic behavior to that of complex **1**, with a very smooth drop while cooling down due to the depopulation of the Stark sublevels, reaching a minimum (21.88 cm^3·K·mol^{-1} for **2**; 18.99 cm^3·K·mol^{-1} for **3**) at 12 K.

The field dependence of the magnetization at 2 K for complexes **1–4** is shown in Figure 2, bottom. For complexes **1–3**, the magnetization increases rapidly below 1 T. For **4**, magnetization presents a value of 3.8 $cm^3 \cdot K \cdot mol^{-1}$, which corresponds to the value of two ferromagnetically coupled Co^{II} cations at 2 K (S = 1/2 for each one). For **1–3**, the values of the magnetization at 5 T are 13.5, 10.9 and 10.9 μ_β, respectively.

The study of the dynamic magnetic properties was also performed for all compounds under a zero magnetic field, revealing a clear dependency of the χ_M'' on temperature and frequency for complex **2** (Figure S3, Supplementary Material), indicating that **2** might be an extremely weak SMM.

3. Materials and Methods

3.1. Materials, Physical and Spectroscopic Measurements

All manipulations were performed under aerobic conditions using materials (reagent grade) and solvents as received. Elemental analyses (C, H, N) were performed by the University of Patras microanalysis service. IR spectra (4000–400 cm^{-1}) were recorded using a Perkin Elmer 16PC FT-IR spectrometer with samples prepared as KBr pellets. Powder X-ray diffraction data (pxrd) were collected using an Inex Equinox 6000 diffractometer. Solid-state, variable-temperature and variable-field magnetic data were collected on powdered samples using an MPMS5 Quantum Design magnetometer operating at 0.03 T in the 300–2.0 K range for the magnetic susceptibility and at 2.0 K in the 0–5 T range for the magnetization measurements. Diamagnetic corrections were applied to the observed susceptibilities using Pascal's constants. Alternating current (*ac*) magnetic susceptibility experiments were carried out at 1000 Hz.

3.2. Synthesis of [Co₂Gd{(py)₂C(OEt)(O)}₄(NO₃)(H₂O)]₂[Gd(NO₃)₅](ClO₄)₂ (1)

Solid $(py)_2CO$ (0.111 g, 0.60 mmol) and $NaO_2CMe \cdot 3H_2O$ (0.041 g, 0.30 mmol) were added to a pink solution of $Co(ClO_4)_2 \cdot 6H_2O$ (0.110 g, 0.30 mmol) in EtOH (15 mL) under stirring, yielding a red solution. $Gd(NO_3)_3 \cdot 6H_2O$ (0.046 g, 0.10 mmol) was then added and the resulting solution was stirred for 30 min. The red solution was allowed to stand undisturbed in a closed flask. Red prismatic crystals appeared after 2 days, which were collected by filtration, washed with EtOH (2 × 5 mL) and Et₂O (2 × 5 mL) and dried in air. Yield: ~65%. Anal. Calcd (Found) for **1**: C, 38.91 (38.80); H, 3.39 (3.72); N, 10.03 (9.73) %. Selected IR data (KBr, cm^{-1}): 3390 (s,b), 2972 (m), 2928 (w), 2897 (w), 1602 (m), 1568 (w), 1470 (s), 1441 (m), 1384 (s), 1317 (s), 1222 (m), 1090 (s), 1053 (s), 903 (w), 777 (m), 686 (m), 635 (m), 624 (m), 541 (w), 474 (m).

3.3. Synthesis of [Co₂Dy{(py)₂C(OEt)(O)}₄(NO₃)(H₂O)]₂[Dy(NO₃)₅](ClO₄)₂ (2)

This was prepared in the same manner as complex **1** but using $Dy(NO_3)_3 \cdot 6H_2O$ (0.046 g, 0.10 mmol) in place of $Gd(NO_3)_3 \cdot 6H_2O$. After 2 days, red prismatic crystals of **2** appeared, which were collected by filtration, washed with EtOH (2 × 5 mL) and Et₂O (2 × 5 mL) and dried in air. Yield: ~60%. Anal. Calcd (Found) for **2**: C, 38.72 (38.91); H, 3.37 (3.75); N, 9.99 (10.08) %. Selected IR data (KBr, cm^{-1}): 3394 (s,b), 2974 (m), 2930 (w), 2897 (w), 1604 (m), 1570 (w), 1472 (s), 1443 (m), 1384 (s), 1315 (s), 1225 (m), 1090 (s), 1054 (s), 904 (w), 780 (m), 686 (m), 635 (m), 625 (m), 542 (w), 472 (m).

3.4. Synthesis of [Co₂Tb{(py)₂C(OEt)(O)}₄(NO₃)(H₂O)]₂[Tb(NO₃)₅](ClO₄)₂ (3)

This was prepared in the same manner as complex **1** but using $Tb(NO_3)_3 \cdot 6H_2O$ (0.046 g, 0.10 mmol) in place of $Gd(NO_3)_3 \cdot 6H_2O$. After 2 days, red prismatic crystals of **3** appeared, which were collected by filtration, washed with EtOH (2 × 5 mL) and Et₂O (2 × 5 mL) and dried in air. Yield: ~65%. Anal. Calcd (Found) for **3**: C, 38.85 (38.73); H, 3.39 (2.99); N, 10.02 (9.84) %. Selected IR data (KBr, cm^{-1}): v = 3394 (s,b), 2974 (m), 2930 (w), 2896 (w), 1604 (m), 1570 (w), 1472 (s), 1442 (m), 1384 (s), 1316 (s), 1224 (m), 1089 (s), 1054 (s), 904 (w), 780 (m), 686 (m), 636 (m), 626 (m), 542 (w), 474 (m).

3.5. Synthesis of [Co$_2$Y{(py)$_2$C(OEt)(O)}$_4$(NO$_3$)(H$_2$O)]$_2$[Y(NO$_3$)$_5$](ClO$_4$)$_2$ (4)

This was prepared in the same manner as complex **1** but using Y(NO$_3$)$_3$·6H$_2$O (0.038 g, 0.10 mmol) in place of Gd(NO$_3$)$_3$·6H$_2$O. After 2 days, red prismatic crystals of **4** appeared, which were collected by filtration, washed with EtOH (2 × 5 mL) and Et$_2$O (2 × 5 mL) and dried in air. Yield: ~65%. Anal. Calcd (Found) for **4**: C, 41.56 (41.47); H, 3.62 (3.53); N, 10.72 (11.09) %. Selected IR data (KBr, cm^{-1}): 3390 (s,b), 2972 (m), 2928 (w), 2897 (w), 1602 (m), 1568 (w), 1470 (s), 1441 (m), 1384 (s), 1317 (s), 1222 (m), 1090 (s), 1053 (s), 903 (w), 777 (m), 686 (m), 635 (m), 624 (m), 541 (w), 474 (m).

Caution! *Although no such behavior was observed during the present work, perchlorate and nitrate salts are potentially explosive; such compounds should be synthesized and used in small quantities, and treated with utmost care at all times.*

3.6. Single-Crystal X-ray Crystallography

Data were collected at the University of Cyprus on an Oxford-Diffraction SuperNova diffractometer, equipped with a CCD area detector and a graphite monochromator utilizing Mo-Kα radiation (λ = 0.71073 Å). Suitable crystals were attached to glass fiber using paratone-N oil and transferred to a goniostat, where they were cooled to 100 K for data collection. Empirical absorption corrections (multi-scan based on symmetry-related measurements) were applied using CrysAlis RED software [51]. The structure was solved by direct methods using SIR92 [52] and refined on F^2 vai the full-matrix least squares method using SHELXL97 [53] and SHELXL-2014/7 [54]. Software packages used are listed as follows: CrysAlisCC for data collection, CrysAlisRED for cell refinement and data reduction [51], WINGX for geometric calculations [55], DIAMOND [56] and MERCURY [57] for molecular graphics. The program SQUEEZE [58], a part of the PLATON package of crystallographic software, was used to remove the contribution of highly disordered solvent molecules. The non-H atoms were treated anisotropically, whereas the H atoms were placed in calculated, ideal positions and refined as riding on their respective C atoms. Unit cell parameters and structure solution and refinement data for **1** are listed in Table S1. An initial search of reciprocal space for **2–4** revealed monoclinic cells with dimensions similar to those of **1**; thus, full data collection of their structures was not pursued.

Several crystals of compound **1**, from different preparations and at different periods of time, were carefully tested on the X-rays (using CuKa and MoK radiation) at ambient and low (100 K) temperatures. The diffraction quality of the crystals proved to be moderate and structure determination was eventually carried out by means of the best data set collected. It is important to mention that compound **1** has a unit cell and structure similar to a Ni2Gd analogous compound, as previously reported by us [36,37], though the latter differs mainly in the nature of the 3d metal ion, i.e., it contains Ni^{II} instead of Co^{II}; thus, although the crystallographic data are not of the best quality, the information they provide about the structure is absolutely reliable.

The X-ray crystallographic data for **1** have been deposited with a CCDC reference number CCDC 1906734. They can be obtained free of charge at http://www.ccdc.cam.ac.uk/conts/retrieving.html or from the Cambridge Crystallographic Data Center, 12 Union Road, Cambridge, CB2 1EZ, UK: Fax: +44-1223-336033; or e-mail: deposit@ccdc.cam.ac.uk.

4. Conclusions

Four new mixed-metal Co$^{II}_2$Ln (Ln = Gd, **1**; Dy, **2**; Tb, **3**) and Co$^{II}_2$Y (**4**) clusters are described, bearing the anionic hemiaketalic form of di-2-pyridyl ketone as an organic ligand. Compounds **1–4** display a triangular metal topology and were synthesized by the reaction of Co(ClO$_4$)$_2$·6H$_2$O, M(NO$_3$)$_3$·6H$_2$O, (py)$_2$CO and CH$_3$CO$_2$Na·3H$_2$O in EtOH. They are the first heterometallic Co/4f or Y clusters containing (py)$_2$CO or its derivatives, and join a very small family of such compounds with this ligand. dc and ac magnetic susceptibility studies revealed the presence of weak ferromagnetic exchange interactions between the metal ions, with **2** exhibiting nonzero, weak out-of-phase (χ"M) signals at temperatures below ~5 K.

(py)$_2$CO remains a rich wellspring of new metal clusters with interesting structural features and magnetic properties, after many years of intense research efforts that have yielded a massive number of compounds. Further studies on the use of this ligand for the synthesis of new 3d/4f metal clusters are in progress and will be reported in due course.

Supplementary Materials: The following are available online at http://www.mdpi.com/2312-7481/5/2/35/s1: Figure S1. Representation of the elipsoid plot for **1**, Figure S2. Theoretical and experimental pxrd patterns for **1–4**, Figure S3: Representation of χ′ (black line) and χ″ (red line) for **2**, Figure S4: Linear fit of the ac magnetic suscetibility data for **2** at the frequency of 1000 Hz using the generalized Debye model to extract the slow relaxation parameters, Table S1: Crystallographic data for complex **1**.

Author Contributions: C.G.E. contributed to the synthesis, crystallization and preliminary characterization of all the compounds, and he wrote the relevant draft of the paper. Á.N.F. contributed to the synthesis of **1–4**. J.M. performed the magnetic measurements, interpreted the results and wrote the relevant part of the paper. A.T. contributed to the structural characterization of **1–4**. S.P.P. contributed to the coordination of the research and interpretation of the results. C.P. contributed to the coordination of the research, collected single crystal X-ray crystallographic data and solved the structure of **1**, and performed refinement in the structure. She also wrote the paper based on the reports of her collaborators.

Funding: J.M. thanks the Ministerio de Economía y Competitividad, Project CTQ2015-63614-P for funding.

Conflicts of Interest: The authors declare no conflict of interest.

References

1. Rosado Piquer, L.; Sañudo, E.C. Heterometallic 3d–4f single-molecule magnets. *Dalton Trans.* **2015**, *44*, 8771–8780. [CrossRef] [PubMed]
2. Liu, K.; Shia, W.; Cheng, P. Toward heterometallic single-molecule magnets: Synthetic strategy, structures and properties of 3d–4f discrete complexes. *Coord. Chem. Rev.* **2015**, *289–290*, 74–122. [CrossRef]
3. Sharples, J.W.; Collison, D. The coordination chemistry and magnetism of some 3d–4f and 4f amino-polyalcohol compounds. *Coord. Chem. Rev.* **2014**, *260*, 1–20. [CrossRef] [PubMed]
4. Chen, W.P.; Liao, P.Q.; Yu, Y.; Zheng, Z.; Chen, X.-M.; Zheng, Y.Z. A Mixed-Ligand Approach for a Gigantic and Hollow Heterometallic Cage {Ni64RE96} for Gas Separation and Magnetic Cooling Applications. *Angew. Chem. Int. Ed.* **2016**, *55*, 9375–9379. [CrossRef] [PubMed]
5. Kong, X.J.; Long, L.-S.; Huang, R.B.; Zheng, L.-S.; Harris, T.D.; Zheng, Z. A four-shell, 136-metal 3d-4f heterometallic cluster approximating a rectangular parallelepiped. *Chem. Commun.* **2009**, 4354–4356. [CrossRef]
6. Kong, X.J.; Ren, Y.P.; Chen, W.X.; Long, L.-S.; Zheng, Z.; Huang, R.-B.; Zheng, L.-S. A Four-Shell, Nesting Doll-like 3d–4f Cluster Containing 108 Metal Ions. *Angew. Chem. Int. Ed.* **2008**, *47*, 2398–2401. [CrossRef]
7. Leng, J.-D.; Liu, J.-L.; Tong, M.-L. Unique nanoscale {CuII36LnIII24} (Ln = Dy and Gd) metallo-rings. *Chem. Commun.* **2012**, *48*, 5286–5288. [CrossRef]
8. Peng, J.-B.; Zhang, Q.-C.; Kong, X.-L.; Zheng, Y.-Z.; Ren, Y.-P.; Long, L.-S.; Huang, R.-B.; Zheng, L.-S.; Zheng, Z. High-Nuclearity 3d−4f Clusters as Enhanced Magnetic Coolers and Molecular Magnets. *J. Am. Chem. Soc.* **2012**, *134*, 3314–3317. [CrossRef]
9. Kong, X.J.; Ren, Y.P.; Long, L.-S.; Zheng, Z.; Nichol, G.; Huang, R.-B.; Zheng, L.-S. Dual Shell-like Magnetic Clusters Containing NiII and LnIII (Ln = La, Pr, and Nd) Ions. *Inorg. Chem.* **2008**, *47*, 2728–2739. [CrossRef]
10. Kong, X.J.; Ren, Y.P.; Long, L.-S.; Zheng, Z.; Huang, R.-B.; Zheng, L.-S. A Keplerate Magnetic Cluster Featuring an Icosidodecahedron of Ni(II) Ions Encapsulating a Dodecahedron of La(III) Ions. *J. Am. Chem. Soc.* **2007**, *129*, 7016–7017. [CrossRef]
11. Bagai, R.; Christou, G. The Drosophila of single-molecule magnetism: [Mn$_{12}$O$_{12}$(O$_2$CR)$_{16}$(H$_2$O)$_4$]. *Chem. Soc. Rev.* **2009**, *38*, 1011–1026. [CrossRef]
12. Christou, G.; Gatteschi, D.; Hendrickson, D.N.; Sessoli, R. Single-Molecule Magnets. *MRS Bull.* **2000**, *25*, 66–71. [CrossRef]
13. Wernsdorfer, W.; Bogani, L. Molecular spintronics using single-molecule magnets. *Nat. Mater.* **2008**, *7*, 179–186.
14. Hill, S.; Edwards, R.S.; Aliaga-Alcalde, N.; Christou, G. Quantum Coherence in an Exchange-Coupled Dimer of Single-Molecule Magnets. *Science* **2003**, *302*, 1015–1018. [CrossRef]

15. Tiron, R.; Wernsdorfer, W.; Aliaga-Alcalde, N.; Christou, G. Quantum tunneling in a three-dimensional network of exchange-coupled single-molecule magnets. *Phys. Rev. B* **2003**, *68*, 140407. [CrossRef]
16. Akhtar, M.N.; Lan, Y.; AlDamen, M.A.; Zheng, Y.-Z.; Anson, C.E.; Powell, A.K. Effect of ligand substitution on the SMM properties of three isostructural families of double-cubane Mn_4Ln_2 coordination clusters. *Dalton Trans.* **2018**, *47*, 3485–3495. [CrossRef]
17. Papatriantafyllopoulou, C.; Wernsdorfer, W.; Abboud, K.A.; Christou, G. $Mn_{21}Dy$ Cluster with a Record Magnetization Reversal Barrier for a Mixed 3d/4f Single-Molecule Magnet. *Inorg. Chem.* **2011**, *50*, 421–423. [CrossRef]
18. Holynska, M.; Premuzic, D.; Jeon, I.-R.; Wernsdorfer, W.; Clérac, R.; Dehnen, S. $[Mn^{III}_6O_3Ln_2]$ Single-Molecule Magnets: Increasing the Energy Barrier Above 100 K. *Chem. Eur. J.* **2011**, *17*, 9605–9610. [CrossRef]
19. Stamatatos, T.C.; Teat, S.J.; Wernsdorfer, W.; Christou, G. Enhancing the Quantum Properties of Manganese-Lanthanide Single-Molecule Magnets: Observation of Quantum Tunneling Steps in the Hysteresis Loops of a {$Mn_{12}Gd$} Cluster. *Angew. Chem. Int. Ed.* **2009**, *48*, 521–524. [CrossRef]
20. Mereacre, V.; Ako, A.M.; Clerac, R.; Wernsdorfer, W.; Filoti, G.; Bartolome, J.; Anson, C.E.; Powell, A.K. A Bell-Shaped $Mn_{11}Gd_2$ Single-Molecule Magnet. *J. Am. Chem. Soc.* **2007**, *129*, 9248–9249. [CrossRef]
21. Mereacre, V.; Ako, A.M.; Clerac, R.; Wernsdorfer, W.; Hewitt, I.J.; Anson, C.E.; Powell, A.K. Heterometallic $[Mn_5$-$Ln_4]$ Single-Molecule Magnets with High Anisotropy Barriers. *Chem. Eur. J.* **2008**, *14*, 3577–3584. [CrossRef]
22. Li, H.; Meng, X.; Wang, M.; Wang, Y.-X.; Shi, W.-S.; Cheng, P. A {Tb_2Fe_3} Pyramid Single-Molecule Magnet with Ferromagnetic Tb-Fe Interaction. *Chin. J. Chem.* **2019**, *37*, 373–377. [CrossRef]
23. Baniodeh, A.; Liang, Y.; Anson, C.E.; Magnani, N.; Powell, A.K.; Unterreiner, A.N.; Seyfferle, S.; Slota, M.; Dressel, M.; Bogani, L.; et al. Unraveling the Influence of Lanthanide Ions on Intra- and Inter-Molecular Electronic Processes in $Fe_{10}Ln_{10}$ Nano-Toruses. *Adv. Funct. Mater.* **2014**, *24*, 6280–6290. [CrossRef]
24. Badia-Romano, L.; Bartolomé, F.; Bartolomé, J.; Luzón, J.; Prodius, D.; Turta, C.; Mereacre, V.; Wilhelm, F.; Rogalev, A. Field-induced internal Fe and Ln spin reorientation in butterfly {Fe_3LnO_2} (Ln = Dy and Gd) single-molecule magnets. *Phys. Rev. B* **2013**, *87*, 184403:1–184403:11. [CrossRef]
25. Schmidt, S.; Prodius, D.; Mereacre, V.; Kostakis, G.E.; Powell, A.K. Unprecedented chemical transformation: Crystallographic evidence for 1,1,2,2-tetrahydroxyethane captured within an Fe_6Dy_3 single molecule magnet. *Chem. Commun.* **2013**, *49*, 1696–1698. [CrossRef]
26. Efthymiou, C.G.; Stamatatos, T.C.; Papatriantafyllopoulou, C.; Tasiopoulos, A.J.; Wernsdorfer, W.; Perlepes, S.P.; Christou, G. Nickel/Lanthanide Single-Molecule Magnets: {Ni_3Ln} "Stars" with a Ligand Derived from the Metal-Promoted Reduction of Di-2-pyridyl Ketone under Solvothermal Conditions. *Inorg. Chem.* **2010**, *49*, 9737–9739. [CrossRef]
27. Zhu, Q.; Xiang, S.; Sheng, T.; Yuan, D.; Shen, C.; Tan, C.; Hua, S.; Wu, X. A series of goblet-like heterometallic pentanuclear $[Ln^{III}Cu^{II}_4]$ clusters featuring ferromagnetic coupling and single-molecule magnet behavior. *Chem. Commun.* **2012**, *48*, 10736–10738. [CrossRef]
28. Ghosh, S.; Ida, Y.; Ishida, T.; Ghosh, A. Linker Stoichiometry-Controlled Stepwise Supramolecular Growth of a Flexible Cu_2Tb Single Molecule Magnet from Monomer to Dimer to One-Dimensional Chain. *Cryst. Growth Des.* **2014**, *14*, 2588–2598. [CrossRef]
29. Jose Heras Ojea, M.; Milway, V.A.; Velmurugan, G.A.; Thomas, L.H.; Coles, S.J.; Wilson, C.; Wernsdorfer, W.; Rajaraman, G.; Murrie, M. Enhancement of Tb^{III}–Cu^{II} Single-Molecule Magnet Performance through Structural Modification. *Chem. Eur. J.* **2016**, *22*, 12839–12848. [CrossRef]
30. Baskar, V.; Gopal, K.; Helliwell, M.; Tuna, F.; Wernsdorfer, W.; Winpenny, R.E.P. 3d–4f Clusters with large spin ground states and SMM behavior. *Dalton Trans.* **2010**, *39*, 4747–4750. [CrossRef]
31. Li, X.-L.; Min, F.-Y.; Wang, C.; Lin, S.-Y.; Liu, Z.; Yang, J. Utilizing 3d-4f magnetic interaction to slow the magnetic relaxation of heterometallic complexes. *Inorg. Chem.* **2015**, *54*, 4337–4344. [CrossRef]
32. Chandrasekhar, V.; Pandian, B.M.; Azhakar, R.; Vittal, J.J.; Clérac, R. Linear Trinuclear Mixed-Metal Co^{II}–Gd^{III}–Co^{II} Single-Molecule Magnet: $[L_2Co_2Gd][NO_3]_2CHCl_3$($LH_3$ = (S)P[N(Me)N=CH–C_6H_3-2-OH-3-OMe]$_3$). *Inorg. Chem.* **2007**, *46*, 5140–5142. [CrossRef]
33. Funes, A.V.; Carrella, L.; Rentschler, E.; Alborés, P. {$Co^{III}_2Dy^{III}_2$} single molecule magnet with two resolved thermal activated magnetization relaxation pathways at zero field. *Dalton Trans.* **2014**, *43*, 2361–2364. [CrossRef]

34. Stamatatos, T.C.; Efthymiou, C.G.; Stoumpos, C.C.; Perlepes, S.P. Adventures in the Coordination Chemistry of Di-2-pyridyl Ketone and Related Ligands: From High-Spin Molecules and Single-Molecule Magnets to Coordination Polymers, and from Structural Aesthetics to an Exciting New Reactivity Chemistry of Coordinated Ligands. *Eur. J. Inorg. Chem.* **2009**, *2009*, 3361–3391.

35. Papaefstathiou, G.S.; Perlepes, S.P. Families of Polynuclear Manganese, Cobalt, Nickel and Copper Complexes Stabilized by Various Forms of Di-2-pyridyl Ketone. *Comments Inorg. Chem.* **2002**, *23*, 249–274. [CrossRef]

36. Efthymiou, C.G.; Georgopoulou, A.N.; Papatriantafyllopoulou, C.; Terzis, A.; Raptopoulou, C.P.; Escuer, A.; Perlepes, S.P. Initial employment of di-2-pyridyl ketone as a route to nickel(II)/lanthanide(III) clusters: Triangular Ni_2Ln complexes. *Dalton Trans.* **2010**, *39*, 8603–8605. [CrossRef]

37. Georgopoulou, A.N.; Efthymiou, C.G.; Papatriantafyllopoulou, C.; Psycharis, V.; Raptopoulou, C.P.; Manos, M.; Tasiopoulos, A.; Escuer, A.; Perlepes, S.P. Triangular $Ni^{II}_2Ln^{III}$ and $Ni^{II}_2Y^{III}$ complexes derived from di-2-pyridyl ketone: Synthesis, structures and magnetic properties. *Polyhedron* **2011**, *30*, 2978–2986. [CrossRef]

38. Papatriantafyllopoulou, C.; Stamatatos, T.C.; Efthymiou, C.G.; Cunha-Silva, L.; Almeida Paz, F.; Perlepes, S.P.; Christou, G. A High-Nuclearity 3d/4f Metal Oxime Cluster: An Unusual Ni_8Dy_8 "Core-Shell" Complex from the Use of 2-Pyridinealdoxime. *Inorg. Chem.* **2010**, *49*, 9743–9745. [CrossRef]

39. Polyzou, C.D.; Efthymiou, C.G.; Escuer, A.; Cunha-Silva, L.; Papatriantafyllopoulou, C.; Perlepes, S.P. In search of 3d/4f-metal single-molecule magnets: Nickel(II)/lanthanide(III) coordination clusters. *Pure Appl. Chem.* **2013**, *85*, 315–327. [CrossRef]

40. Papatriantafyllopoulou, C.; Estrader, M.; Efthymiou, C.G.; Dermitzaki, D.; Gkotsis, K.; Terzis, A.; Diaz, C.; Perlepes, S.P. In search for mixed transition metal/lanthanide single-molecule magnets: Synthetic routes to Ni^{II}/Tb^{III} and Ni^{II}/Dy^{III} clusters featuring a 2-pyridyl oximate ligand. *Polyhedron* **2009**, *28*, 1652–1655. [CrossRef]

41. Savva, M.; Skordi, K.; Fournet, A.D.; Thuijs, A.E.; Christou, G.; Perlepes, S.P.; Papatriantafyllopoulou, C.; Tasiopoulos, A.J. Heterometallic $Mn^{III}_4Ln_2$ (Ln = Dy, Gd, Tb) Cross-Shaped Clusters and Their Homometallic $Mn^{III}_4Mn^{II}_2$ Analogues. *Inorg. Chem.* **2017**, *56*, 5657–5668. [CrossRef]

42. Liu, W.; Thorp, H.H. Bond Valence Sum Analysis of Metal-Ligand Bond Lengths in Metalloenzymes and Model Complexes. 2. Refined Distances and Other Enzymes. *Inorg. Chem.* **1993**, *32*, 4102–4105. [CrossRef]

43. Ruiz-Martinez, A.; Casanova, D.; Alvarez, S. Polyhedral Structures with an Odd Number of Vertices: Nine-Coordinate Metal Compounds. *Chem. Eur. J.* **2008**, *14*, 1291–1303. [CrossRef]

44. Llunell, M.; Casanova, D.; Girera, J.; Alemany, P.; Alvarez, S. *SHAPE*, version 2.0; Universitat de Barcelona: Barcelona, Spain, 2010.

45. Georgopoulou, A.N.; Adam, R.; Raptopoulou, C.P.; Psycharis, V.; Ballesteros, R.; Abarca, B.; Boudalis, A.K. Expanding the 3d-4f heterometallic chemistry of the $(py)_2CO$ and pyCOpyCOpy ligands: Structural, magnetic and Mössbauer spectroscopic studies of two Fe^{II}–Gd^{III} complexes. *Dalton Trans.* **2011**, *40*, 8199–8205. [CrossRef]

46. Alexandropoulos, D.I.; Cunha-Silva, L.; Tang, J.; Stamatatos, T.C. Heterometallic Cu/Ln cluster chemistry: Ferromagnetically-coupled $\{Cu_4Ln_2\}$ complexes exhibiting single-molecule magnetism and magnetocaloric properties. *Dalton Trans.* **2018**, *47*, 11934–11941. [CrossRef]

47. Tang, J.; Zhang, P. *Lanthanide Single Molecule Magnets*; Springer: Berlin/Heidelberg, Germany, 2015.

48. Lloret, F.; Julve, M.; Cano, J.; Ruiz-García, R.; Pardo, E. Magnetic properties of six-coordinated high-spin cobalt(II) complexes: Theoretical background and its application. *Inorg. Chim. Acta* **2008**, *361*, 3432–3445. [CrossRef]

49. Xu, Y.; Luo, F.; Zheng, J.-M. Syntheses, Structures, and Magnetic Properties of a Series of Heterotri-, Tetra- and Pentanuclear Ln^{III}–Co^{II} Compounds. *Polymers* **2019**, *11*, 196. [CrossRef]

50. Bartolomé, J.; Filoti, G.; Kuncser, V.; Schinteie, G.; Mereacre, V.; Anson, C.E.; Powell, A.K.; Prodius, D.; Turta, C. Magnetostructural correlations in the tetranuclear series of $\{Fe_3LnO_2\}$ butterfly core clusters:Magnetic and Mössbauer spectroscopic study. *Phys. Rev. B* **2009**, *80*, 014430:1–014430:16. [CrossRef]

51. Oxford Diffraction. *CrysAlis CCD and CrysAlis RED*, version 1.171.32.15; Oxford Diffraction Ltd.: Oxford, UK, 2008.

52. Altomare, A.; Cascarano, G.; Giaconazzo, C.; Guagliardi, A.; Burla, M.C.; Polidori, G.; Camalli, M. *SIR92: A program for automatic solution of crystal structures by direct methods. J. Appl. Crystallogr.* **1994**, *27*, 435–436. [CrossRef]

53. Sheldrick, G.M. *SHELXL97*; University of Göttingen: Göttingen, Germany, 1997.
54. Sheldrick, G.M. *SHELXL-2014/7*; Program for Refinement of Crystal Structures, University of Göttingen: Göttingen, Germany, 2014.
55. Farrugia, L.J. WinGX suite for small-molecule single-crystal crystallography. *J. Appl. Crystallogr.* **1999**, *32*, 837–838. [CrossRef]
56. Brandenburg, K. *DIAMOND*; Version 3.1d; Crystal Impact GbR: Bonn, Germany, 2006.
57. Macrae, C.F.; Edgington, P.R.; McCabe, P.; Pidcock, E.; Shields, G.P.; Taylor, R.; Towler, M.; Van de Streek, J. Mercury: Visualization and analysis of crystal structures. *J. Appl. Crystallogr.* **2006**, *39*, 453–457. [CrossRef]
58. Van der Sluis, P.; Spek, A.L. BYPASS: An Effective Method for the Refinement of Crystal Structures Containing Disordered Solvent Regions. *Acta Crystallogr. Sect. A Found Crystallogr.* **1990**, *A46*, 194–201. [CrossRef]

 *magnetochemistry*

Article

Diversity of Coordination Modes in a Flexible Ditopic Ligand Containing 2-Pyridyl, Carbonyl and Hydrazone Functionalities: Mononuclear and Dinuclear Cobalt(III) Complexes, and Tetranuclear Copper(II) and Nickel(II) Clusters [†]

Evangelos Pilichos [1], Evangelos Spanakis [1], Evangelia-Konstantina Maniaki [1], Catherine P. Raptopoulou [2], Vassilis Psycharis [2,*], Mark M. Turnbull [3,*] and Spyros P. Perlepes [1,4,*]

[1] Department of Chemistry, University of Patras, 26504 Patras, Greece
[2] Institute of Nanoscience and Nanotechnology, NCSR "Demokritos", 15310 Aghia Paraskevi Attikis, Greece
[3] Carlson School of Chemistry and Biochemistry, Clark University, Worcester, MA 01610, USA
[4] Institute of Chemical Engineering Sciences, Foundation for Research and Technology-Hellas (FORTH/ICE-HT), Platani, B.O. Box 1414, 26504 Patras, Greece
[*] Correspondence: v.psycharis@inn.demokritos.gr (V.P.); MTurnbull@clarku.edu (M.M.T.); perlepes@patreas.upatras.gr (S.P.P.); Tel.: +30-210-6503346 (V.P.); +1-508-7937167 (M.M.T.); +30-2610-996730 (S.P.P.)
[†] This article is dedicated to Professor Masahiro Yamashita—a great scientist and a precious friend—on the occasion of his 65th birthday.

Received: 15 May 2019; Accepted: 12 June 2019; Published: 1 July 2019

Abstract: Syntheses, crystal structures and characterization are reported for four new complexes $[Cu_4Br_2(L)_4]Br_2$ (**1**), $[Ni_4(NO_3)_2(L)_4(H_2O)](NO_3)_2$ (**2**), $[Co_2(L)_3](ClO_4)_3$ (**3**) and $[Co(L)_2](ClO_4)$ (**4**), where L^- is the monoanion of the ditopic ligand N'-(1-(pyridin-2-yl)ethylidene)pyridine-2-carbohydrazide (LH) built on a picolinoyl hydrazone core fragment, and possessing a bidentate and a tridentate coordination pocket. The tetranuclear cation of **1**·0.8H$_2$O·MeOH is a strictly planar, rectangular [2 × 2] grid. Two 2.21011 L^- ligands bridge adjacent Cu^{II} atoms on the short sides of the rectangle through their alkoxide oxygen atoms, and two 2.11111 ligands bridge adjacent Cu^{II} atoms on the long sides of the rectangle through their diazine groups; two metal ions are 5-coordinate and two are 6-coordinate. The tetranuclear cation of **2**·0.2H$_2$O·3EtOH is a square [2 × 2] grid. The two 6-coordinate Ni^{II} atoms of each side of the square are bridged by the alkoxide O atom of one 2.21011 L^- ligand. The dinuclear cation of **3**·0.8H$_2$O·1.3MeOH contains two low-spin octahedral Co^{III} ions bridged by three 2.01111 L^- ligands forming a *pseudo* triple helicate. In the mononuclear cation $[Co(L)_2]^+$ of complex **4**, the low-spin octahedral Co^{III} center is coordinated by two tridentate chelating, meridional 1.10011 ligands. The crystal structures of the complexes are stabilized by a variety of π–π stacking and/or H-bonding interactions. Compounds **2**, **3** and **4** are the first structurally characterized nickel and cobalt complexes of any form (neutral or anionic) of LH. The 2.01111 and 1.10011 coordination modes of L^-, observed in the structures of complexes **3** and **4**, have been crystallographically established for the first time in coordination complexes containing this anionic ligand. Variable-temperature, solid-state dc magnetic susceptibility and variable-field magnetization studies at 1.8 K were carried out on complexes **1** and **2**. Antiferromagnetic metal ion···metal ion exchange interactions are present in both complexes. The study reveals that the cation of **1** can be considered as a practically isolated pair of strongly antiferromagnetically coupled (through the diazine group of L^-) dinulear units. The susceptibility data for **2** were fit to a single-J model for an $S = 1$ cyclic tetramer. The values of the J parameters have been rationalized in terms of known magnetostructural correlations. Spectral data (infrared (IR), ultraviolet/visible (UV/VIS), ^{1}H nuclear magnetic resonance (NMR) for the diamagnetic complexes) are also discussed in the light of the structural features of **1–4** and the coordination modes of the organic and inorganic ligands that are present in the complexes. The combined work demonstrates the ligating flexibility of L^-, and its usefulness in the synthesis of complexes with interesting structures and properties.

Keywords: coordination clusters; [2 × 2] grids; magnetic studies; N'-(1-(pyridin-2-yl)ethylidene) pyridine-2-carbohydrazide cobalt(III); nickel(II) and copper(II) complexes

---

1. Introduction

The word "ligand" is derived from the Latin verb *"ligare"* meaning "to bind" [1]. It was first introduced by Alfred Stock when lecturing in Berlin (1916) on the chemistry of boranes and silanes. However, it came into common use only in the 1950s, mainly through the PhD Thesis of Jannik Bjerrum [2]. Nowadays, the appropriate use of old ligands and the design of new, sophisticated ones is one of the pylons of modern inorganic chemistry. Theoretical concepts related to ligands include the chelate effect, the macrocyclic effect, the conformation of chelating rings, the chemistry of non-innocent ligands, the hard and soft bases concept and the isoelectronic and isolobal relationships, among others. Of particular interest is also the study of the reactivity of coordinated ligands, an approach in which the metal ion activates a proligand, transforming it through an in situ reaction and providing unusual ligands that sometimes cannot be synthesized by conventional organic or inorganic synthesis [3,4]. The proper choice of bridging ligands has played a key role in the development of modern magnetochemistry and the interdisciplinary field of molecular magnetism [5], where the metal···metal exchange interactions mediated through the bridges are responsible for a variety of interesting magnetic phenomena [6–9].

Polytopic organic ligands are particularly interesting in coordination chemistry and magnetochemistry. Their design and subsequent synthesis introduces preprogrammed coordination information that is "stored" in the coordination pockets [10,11]. When such ligands react with a transition metal ion, it interprets this information according to its own coordination "algorithm". If the coordination pocket does not contain many donor atoms to fully saturate the coordination requirements of the metal ion, self-assembly can take place favoring the formation of homoleptic or heteroleptic coordination clusters [12]. The obtained nuclearity depends largely on the polytopic nature of the ligand and the preferred metal ion coordination number and geometry [10–13]. This in turn leads to a wide variety of magnetic exchange interactions, which depend on the number and nature of bridges and the magnetic orbitals that are available.

The ligand of the present work is N'-(1-(pyridin-2-yl)ethylidene)pyridine-2-carbohydrazide [other names: methyl(pyridin-2-yl)methanone picolinoylhydrazone or 2-acetylpyridine picolinoylhydrazone], drawn in its enol-imino form in Scheme 1 and abbreviated as LH. It is a ditopic ligand built on a picolinoyl hydrazone core fragment (it can also be considered as an asymmetric alkoxy diazine ligand [14]) possessing a bidentate and a tridentate coordination pocket. The deprotonated ligand (L^-) has two potentially bridging functional groups (μ-O, μ-N-N) and, because of the free rotation around the N–N single bond, can exist in two different coordination conformers, both of which can in principle form spin-coupled dinuclear and polynuclear metal complexes with quite different magnetic properties. We decided to work with this ligand because its published coordination chemistry has been limited [14–21]. Since no Co(II) and Ni(II) complexes of L^- have been reported, we first targeted compounds with these metal ions. We were also interested in preparing Cu(II) complexes, because the only reported complex $[Cu_4(L)_4(H_2O)_2](NO_3)_4$ [14] is a structurally impressive square [2 × 2] grid and can be considered magnetically as an essentially isolated pair of antiferromagnetically coupled dinuclear fragments. We report herein our results from the synthetic investigation of the $CuBr_2/LH$, $Ni(NO_3)_2 \cdot 6H_2O/LH$ and $Co(ClO_4)_2 \cdot 6H_2O/LH$ reaction systems and the characterization of the products obtained. This paper can be considered as a continuation of our interest in the chemistry and magnetism of 3d-metal coordination clusters [9,22], and in the coordination and metal ion-meditated/promoted transformation properties of polydentate ligands containing two or more functionalities (including 2-pyridyl, carbonyl and hydrazone/azine groups, among others) [4,23–27].

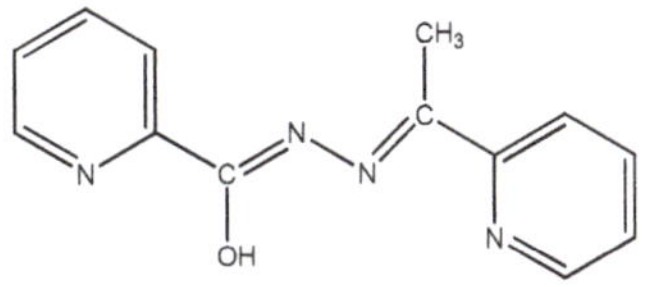

LH

Scheme 1. The free ligand N'-(1-(pyridin-2-yl)ethylidene)pyridine-2-carbohydrazide (LH) drawn in its enol-imino form.

2. Results and Discussion

2.1. Synthetic Comments

A variety of $M^{II}/X^-/LH/B$ (M = Co, Ni, Cu; X = Cl, Br, NO_3, ClO_4; B = Et_3N, LiOH, R_4NOH, NaO_2CR' with R,R' = various groups) reaction systems, involving various solvent media, reagent ratios and crystallization techniques, were systematically investigated before arriving at the optimized synthetic conditions reported in Section 3. In many instances we have isolated microcrystalline powders with reasonable analytical data, but we report here only the structurally characterized products.

The $CuBr_2/NaO_2CPh/LH$ (1:1:1) reaction mixture in MeOH gave a green solution from which greenish brown crystals of $[Cu_4Br_2(L)_4]Br_2 \cdot 0.8H_2O \cdot MeOH$ (1·0.8H_2O·MeOH) were subsequently isolated in a good yield (~60%). Assuming that **1** is the only product from the reaction system, its formation can be summarized by Equation (1). Use of other bases, e.g., Et_3N and $Me_4NOH \cdot 5H_2O$, gave powders of the same product (infrared (IR) evidence).

$$4\,CuBr_2 + 4\,LH + 4\,NaO_2CPh \xrightarrow{MeOH} [Cu_4Br_2(L)_4]Br_2 + 4\,PhCO_2H + 4\,NaBr \qquad (1)$$

Complex $[Ni_4(NO_3)_2(L)_4(H_2O)](NO_3)_2$ (**2**), crystallographically characterized as $2 \cdot 0.2H_2O \cdot 3EtOH$, was prepared by the 1:1 reaction between $Ni(NO_3)_2 \cdot 6H_2O$ and LH in CH_2Cl_2-EtOH, Equation (2), in a rather low yield (~30%). The use of CH_2Cl_2 was necessary to improve the quality of the obtained brown crystals. Use of Et_3N in the reaction mixture gave the same complex in a powder form, Equation (3), but—somewhat to our surprise—with no significant yield improvement.

$$4\,Ni(NO_3)_2\,6H_2O + 4\,LH \xrightarrow{EtOH-CH_2Cl_2} [Ni_4(NO_3)_2(L)_4(H_2O)](NO_3)_2 + 4\,HNO_3 + 23\,H_2O \quad (2)$$

$$4\,Ni(NO_3)_2\,6H_2O + 4\,LH + 4\,Et_3N \xrightarrow{EtOH-CH_2Cl_2} [Ni_4(NO_3)_2(L)_4(H_2O)](NO_3)_2 + 4\,(Et_3NH)(NO_3) + 23\,H_2O \quad (3)$$

Depending on the Co(II): LH reaction ratio used, the $Co(ClO_4)_2 \cdot 6H_2O/LH$ reaction system gave two products in MeOH *under aerobic conditions*, namely $[Co_2(L)_3](ClO_4)_3$ (**3**), crystallographically formulated as $3 \cdot 0.8H_2O \cdot 1.3MeOH$, and $[Co(L)_2](ClO_4)$ (**4**) in moderate yields (~50%). Both products are Co(III) complexes, the atmospheric air oxygen being the oxidant; the oxidation is certainly facilitated by the N-rich environment from the ligand. The 2:3 reaction between $Co(ClO_4)_2 \cdot 6H_2O$ and LH gives complex **3** according to Equation (4). The addition of the base is not necessary for the formation and isolation of the dinuclear complex; its presence increases slightly the yield of the reaction. Use of an excess of LH (LH: Co^{II} = 2:1) has provided access to the 1:2 mononuclear cationic complex 4; the use of base here is necessary for the isolation of the compound in satisfactory yields, Equation (5). Complex **3** can be converted to compound **4** (albeit in a low yield) in MeOH under reflux, Equation (6). The yield can be impressively improved by the addition of base, e.g., LiOH, Equation (7).

$$4\,Co^{II}(ClO_4)_2\cdot 6H_2O + 6\,LH + O_2 \xrightarrow{MeOH} 2\,[Co^{III,III}_2\,(L)_3](ClO_4)_3 + 2\,HClO_4 + 26\,H_2O \qquad (4)$$

$$4\,Co^{II}(ClO_4)_2\cdot 6H_2O + 8\,LH + 4\,Et_3N + O_2 \xrightarrow{MeOH} 4\,[Co^{III}(L)_2](ClO_4) + 4\,(Et_3NH)(ClO_4) + 26\,H_2O \qquad (5)$$

$$[Co_2(L)_3](ClO_4)_3 + LH \xrightarrow[T]{MeOH} 2\,[Co\,(L)_2](ClO_4) + HClO_4 \qquad (6)$$

$$[Co_2(L)_3](ClO_4)_3 + LH + LiOH \xrightarrow[T]{MeOH} 2\,[Co(L)_2](ClO_4) + LiClO_4 + H_2O \qquad (7)$$

2.2. Spectroscopic Characterization in Brief

IR and ultraviolet/visible (UV/VIS) spectra of the complexes were obtained from analytically pure samples which have the formulae **1**, **2**, **3**·H_2O and **4** (Section 3). In the IR spectrum of sample **3**·H_2O, the broad band centered at ~3420 cm^{-1} is due to the v(OH) vibration of the lattice water [7]. The v(OH) vibration of coordinated H_2O in the spectrum of **2** also appears in this region. The IR spectrum of the free ligand LH exhibits a medium-intensity band at 3316 cm^{-1} and a very strong band at 1702 cm^{-1}, assigned to the v(NH) and v(C = O) vibrations, respectively [14,15]. The appearance of these stretching vibrations indicates that LH is present in its keto-amino form, and not in the enol-imino form drawn in Figure 1. Such vibrations are absent from the spectra of the complexes, the spectral regions 3400–3100 cm^{-1} and 1700–1610 cm^{-1} showing no bands. The absence of these bands indicates that (i) the ligands are deprotonated in the complexes and (ii) the carbon–oxygen bond of coordinated L^- does not have an appreciable double bond character [23]; these facts are confirmed in the crystal structures of the complexes (*vide infra*). The highest wavenumber bands in the 2000–400 cm^{-1} region are at 1594 (**1**), 1598 (**2**), 1606 (**3** H_2O) and 1602 (**4**) cm^{-1}, assigned to a pyridyl stretching vibration [7].

The KBr spectrum of **2** exhibits a strong sharp band at 1384 cm^{-1}, assigned to the v_3(E′)[v_d(NO)] vibrational mode of the planar ionic nitrate of D_{3h} symmetry [28]. The absence of bands that would be indicative of the monodentate and bidentate coordinated nitrato groups (present in the structure of the cluster) is rather surprising. This suggests [29,30] that the nitrato ligands are replaced by bromides that are in excess in the KBr matrix, thus producing ionic nitrates (KNO_3); this replacement is facilitated by the pressure that is applied for the preparation of the KBr matrix. As expected, extra nitrato bands appear in the mull (nujol, hexachlorobutadiene) spectra of **2**. For example, the bands at 1501 and 1300 cm^{-1} are assigned [28–30] to the v_1(A$_1$)[v(N = O)] and v_5(B$_2$)[v_{as}(NO$_2$)] vibrational modes, respectively, of the coordinated nitrato group. The separation of these two bands is large (~200 cm^{-1}), suggesting a bidentate nitrato ligand of C_{2v} symmetry [28]. The v_5(B$_2$)[v_{as}(NO$_2$)] band of the monodentate nitrato group could not be assigned with certainty in the mull spectra because other bands of stretching vibrations origin appear in the 1450–1350 cm^{-1} region. The band at 1294 cm^{-1} is a serious candidate for the v_1(A$_1$)[v_s(NO$_2$)] vibration of the monodentate nitrato ligand which is expected around 1300 cm^{-1} [28]. The spectra of **3**·H_2O and **4** exhibit a strong band at 1090–1080 and a medium-intensity band at ~625 cm^{-1}, attributable to the IR-active v_3(F$_2$)[v_d(Cl-O)] and v_4(F$_2$)[δ_d(OClO)] vibrations of the uncoordinated T$_d$ ClO$_4^-$ counterion, respectively [28].

The d-d spectrum of **1** in MeOH consists of a featureless band at 745 nm; this wavelength is fairly typical of a distorted square pyramidal or/and a tetragonally distorted six-coordinate geometry [31,32]. The spectrum also exhibits an absorption at ~370 nm assignable to a Br$^-$-to Cu^{II} LMCT transition [32]. Copper(II) is relatively easy to reduce to copper(I) and the observed transition from a π orbital of the bromo ligand to the singly occupied 3d orbital of Cu^{II} occurs at a relatively low energy (~27,000 cm^{-1}). The d-d spectrum of **2** in MeOH consists of three bands at 365, 615 and 980 nm assignable [32,33] to the $^3A_{2g} \rightarrow{}^3T_{1g}$ (P), $^3A_{2g} \rightarrow{}^3T_{1g}$ (F) and $^3A_{2g} \rightarrow{}^3T_{2g}$ transitions, respectively, in an octahedral 3d^8 ligand field; the wavelengths are typical of Ni(II) chromophores possessing both N and O donors [32,33]. The UV/VIS spectra of concentrated solutions of **3**·H_2O and **4** in MeCN are typical for low-spin octahedral

{Co^{III}N$_6$} and {Co^{III}N$_x$O$_{6-x}$} chromophores, respectively [32,34]. The low-spin octahedral ground term is $^1A_{1g}$ and there are two spin-allowed transitions, with lower lying spin triplet partners, all derived from $(t_{2g})^5(e_g)^1$. Under this scheme, the bands/shoulders at 395, 440, 580 and 735 nm in the spectrum of **3**·H$_2$O are assigned to the $^1A_{1g} \rightarrow {}^1T_{2g}$, $^1A_{1g} \rightarrow {}^1T_{1g}$, $^1A_{1g} \rightarrow {}^3T_{2g}$ and $^1A_{1g} \rightarrow {}^3T_{1g}$, respectively. The corresponding transitions in the spectrum of **4** appear at 405, 450, 590 and 760 nm.

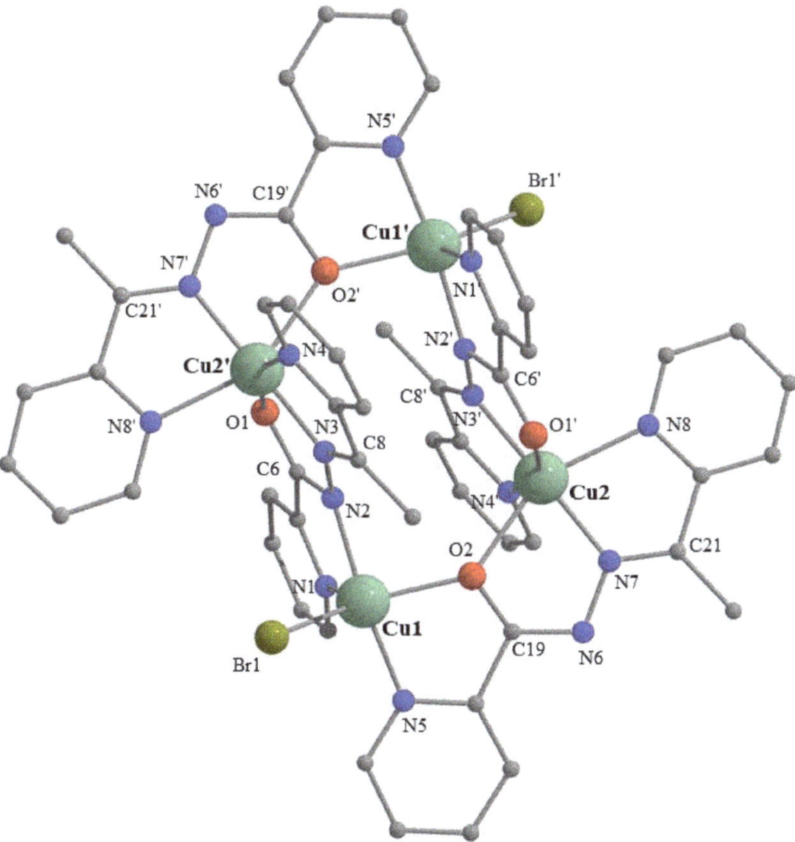

Figure 1. Partially labelled plot of the structure of the cation [Cu$_4$Br$_2$(L)$_4$]$^{+2}$ that is present in the crystal. Structure of **1**·0.8H$_2$O·MeOH. Symmetry code: (′) = −x + 1, −y + 2, −z. A plot with thermal ellipsoids is presented in Figure S17.

The ^{1}H nuclear magnetic resonance (NMR) spectra of the diamagnetic, analytically pure samples **3**·H$_2$O and **4** in deuterated dimethyl sulfoxide (DMSO-d$_6$) are almost identical, except the extra peak at δ 3.17 ppm in the former due to the protons of the lattice H$_2$O. Singlet resonances of the methyl protons were observed at the rather low-field δ value of 3.35 ppm, while the resonances of the eight pyridyl rings appear in the δ region 8.60–7.45 ppm; the integration ratio is 3:8, as expected. These facts indicate the presence of a single L$^-$ species in solution, but our data—combined with literature reports [18]—do not permit safe conclusions concerning the coordinated or non-coordinated nature of the anionic species in solution.

Representative spectra of the complexes are presented in Figures S1–S16.

2.3. Description of Structures

The structures of the four complexes have been solved by single-crystal, X-ray crystallography. Aspects of the molecular and crystal structures are shown in Figures 1–8 and Figures S17–S21. Crystallographic data are presented in Table S1, while numerical data concerning interatomic distances, bond angles and H-bonding interactions are listed in Tables 1–4 and Figures S2–S5.

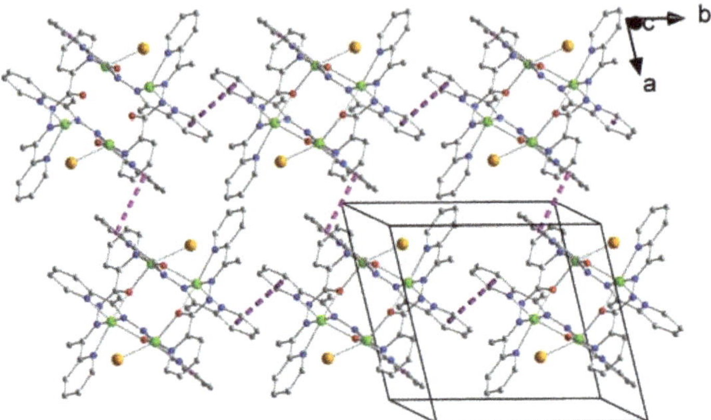

Figure 2. A layer of the tetranuclear cations of complex **1**·0.8H$_2$O·MeOH parallel to the (001) plane. The dashed pink and violet lines indicate π–π stacking interactions between centrosymmetrically-related pairs of aromatic rings containing the N1 and N4 atoms, respectively.

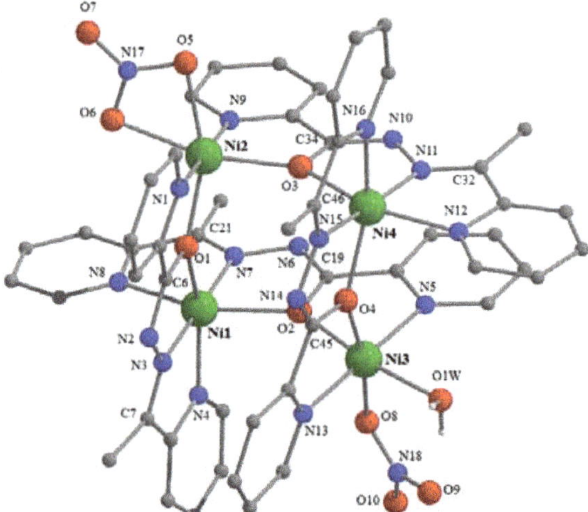

Figure 3. Partially labelled plot of the structure of the cation [Ni$_4$(NO$_3$)$_2$(L)$_4$(H$_2$O)]$^{2+}$ that is present in the.crystal structure of **2**·0.2H$_2$O·3EtOH. A plot with thermal ellipsoids is presented in Figure S19.

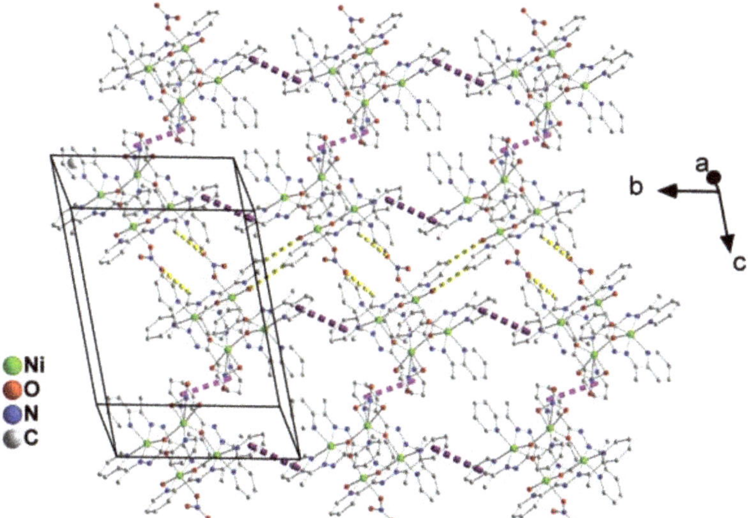

Figure 4. A layer of the tetranuclear cations of complex **2**·0.2H$_2$O·3EtOH parallel to the (100) plane. The dashed pink and violet lines indicate interchain and intrachain, respectively, π–π stacking interactions. The dashed yellow lines represent interstripe H-bonding interactions.

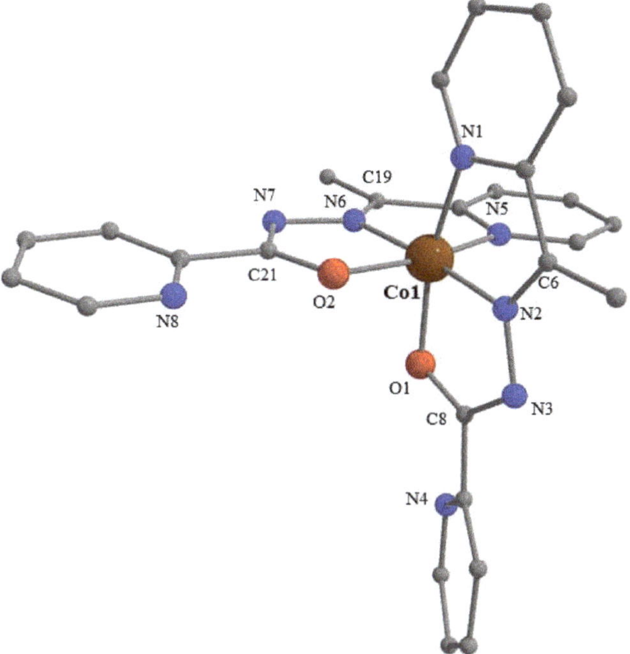

Figure 5. Partially labelled plot of the structure of the cation [Co(L)$_2$]$^+$ that is present in the crystal structure of **4**. A plot with thermal ellipsoids is presented in Figure S20.

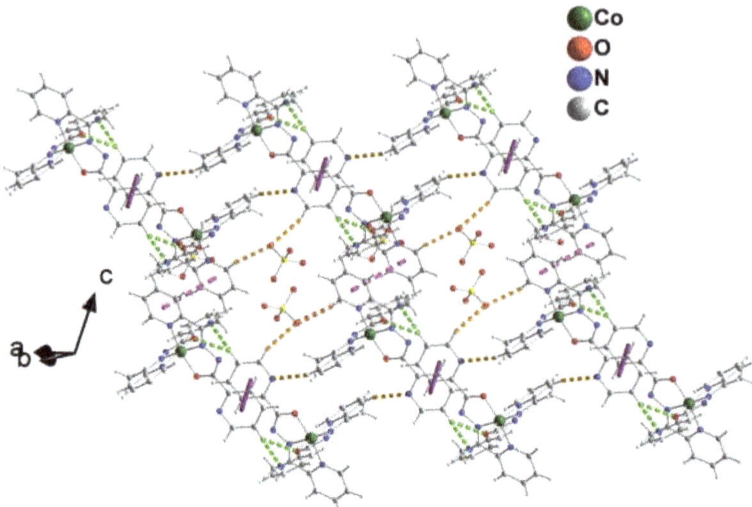

Figure 6. A layer of **4** parallel to the (1–10) plane. The dashed pink and solid violet lines indicate π–π. interactions between pairs of N1- and N4-containing rings, respectively. The dashed green lines represent the C12-H(C12)$\cdots$N8 and C12-H(C12)$\cdots$O2 intrachain H bonds. The dashed orange lines represent the C17-H(C17)$\cdots$N4, C1-H(C1)$\cdots$O5 and C13-H(C13)$\cdots$O5 interchain H bonds. Atoms C1, C12, C13 and C17 are aromatic carbon atoms not labelled in Figure 5 and Figure S4. Atom O5 belongs to the ClO_4^- counterion not shown in Figure 5 and Figure S4.

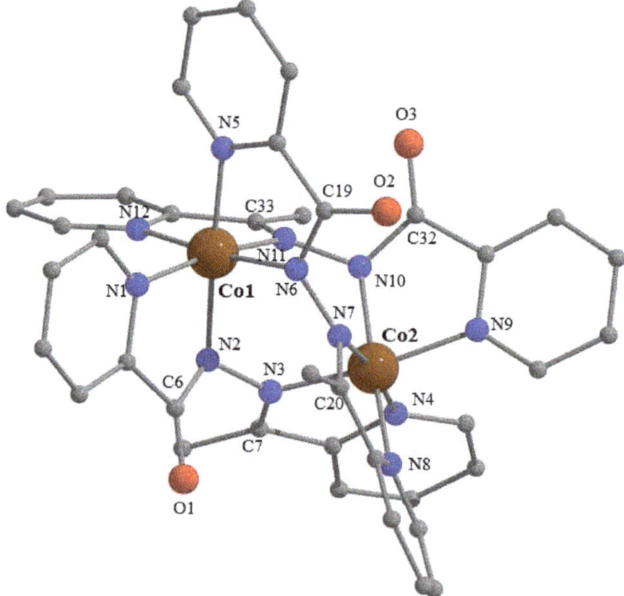

Figure 7. Partially labelled plot of the cation $[Co_2(L)_3]^{3+}$ that is present in the crystal structure of **3**·0.8H$_2$O·1.3MeOH. A plot with thermal ellipsoids is presented in Figure S21 (left).

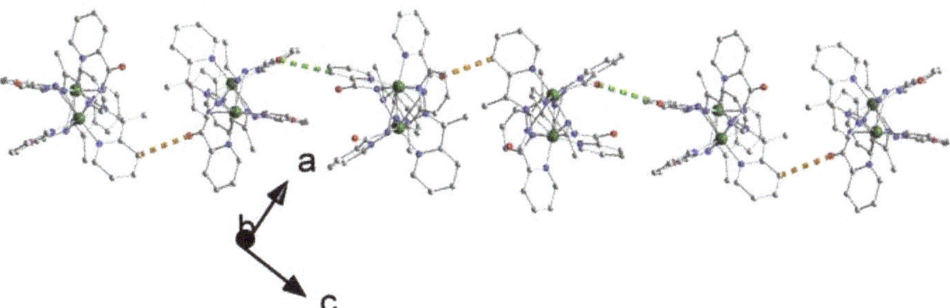

Figure 8. A chain of $[Co_2(L)_3]^{3+}$ cations parallel to the [101] direction in the crystal structure of **3**·0.8H$_2$O·1.3MeOH. The dashed orange and green lines indicate the H-bonding interactions C9-H(C9)···O1 and C17-H(C17)···O3, respectively. Atoms C9 and C17 are aromatic carbon atoms, not labelled in Figure 7 and Figure S21.

Table 1. Selected interatomic distances (Å) and bond angles (°) for complex **1**·0.8H$_2$O·MeOH [a].

Interatomic Distances (Å)		Bond Angles (°)	
Cu1-N1	2.143(5)	N2-Cu1-N5	172.7(2)
Cu1-N2	1.989(5)	O2-Cu1-Br1	147.7(1)
Cu1-O2	2.073(4)	N1-Cu1-Br1	117.0(1)
Cu1-N5	1.994(5)	N1-Cu1-N2	80.3(2)
Cu1-Br1	2.450(1)	N1-Cu1-N5	97.0(2)
Cu2-N7	1.980(5)	N1-Cu1-O2	95.3(2)
Cu2-N8	2.193(5)	N2-Cu1-O2	93.4(2)
Cu2-O2	2.331(4)	N5-Cu1-Br1	95.1(1)
Cu2-N3′	1.975(5)	N3′-Cu2-N7	175.5(2)
Cu2-N4′	2.054(5)	N8-Cu2-O2	148.2(2)
Cu2-O1′	2.117(4)	N4′-Cu2-O1′	155.4(2)
C6-O1	1.257(7)	N3′-Cu2-N4′	79.1(2)
C6-N2	1.346(8)	N7-Cu2-N8	77.4(2)
N2-N3	1.384(7)	O2-C19-N6	126.8(5)
N3-C8	1.290(8)	C19-N6-N7	110.3(5)
C19-O2	1.287(7)	N6-N7-C21	118.1(5)
C19-N6	1.331(7)	O1-C6-N2	125.0(5)
N6-N7	1.401(6)	C6-N2-N3	110.2(5)
N7-C21	1.281(7)	N2-N3-C8	123.7(5)

[a] Symmetry code (′) = −x + 1, −y + 2, −z.

The crystal structure of **1**·0.8H$_2$O·MeOH consists of tetranuclear cations $[Cu_4Br_2(L)_4]^{+2}$ (Figure 1 and Figure S17), Br$^-$ counterions, and lattice H$_2$O and MeOH molecules. The cation possesses a crystallographic inversion center in the midpoint of the Cu1···Cu1′ (or Cu2···Cu2′) distance. The cation is a rectangular [2 × 2] grid; all the metal centers are strictly coplanar (by symmetry). The sides of the rectangle are 4.121(1) Å (Cu1···Cu2/Cu1′···Cu2′) and 4.797(1) Å (Cu1···Cu2′/Cu1′···Cu2), and the diagonals are 6.690(1) Å (Cu1···Cu1′) and 5.935(1) Å (Cu2···Cu2′).

Table 2. Selected interatomic distances (Å) and bond angles (°) for complex 2·0.2H$_2$O·3EtOH.

Interatomic Distances (Å)		Bond Angles (°)	
Ni1···Ni2	3.922(1)	O1-Ni1-N4	154.1(1)
Ni1···N3	3.938(1)	O2-Ni1-N8	153.9(1)
Ni1···N4	5.526(1)	N3-Ni1-N7	177.9(1)
Ni2···Ni3	5.559(1)	O1-Ni1-O2	91.4(1)
Ni2···Ni4	3.921(1)	O2-Ni1-N3	105.3(1)
Ni3···Ni4	3.895(1)	N3-Ni1-N4	78.0(1)
Ni1-O1	2.147(2)	O1-Ni2-O5	158.2(1)
Ni1-O2	2.144(2)	O3-Ni2-O6	165.1(1)
Ni1-N3	1.988(2)	N1-Ni2-N9	177.5(1)
Ni1-N4	2.083(2)	O1-Ni2-O3	95.0(1)
Ni1-N7	1.984(2)	O5-Ni2-O6	60.7(1)
Ni1-N8	2.107(2)	O6-Ni2-N9	93.3(1)
Ni2-O1	2.049(2)	O2-Ni3-O1W	169.9(1)
Ni2-O3	2.054(2)	O4-Ni3-O8	171.1(1)
Ni2-O5	2.161(2)	N5-Ni3-N13	179.1(1)
Ni2-O6	2.113(2)	O2-Ni3-O4	90.5(1)
Ni2-N1	2.041(2)	O1W-Ni3-O8	93.9(1)
Ni2-N9	2.036(2)	O8-Ni3-N5	86.8(1)
Ni3-O2	2.068(2)	O3-Ni4-N12	154.5(1)
Ni3-O4	2.052(2)	O4-Ni4-N16	154.7(1)
Ni3-O8	2.121(2)	N11-Ni4-N15	174.5(1)
Ni3-O1W	2.080(2)	O3-Ni4-O4	94.4(1)
Ni3-N5	2.048(2)	N11-Ni4-N12	78.5(1)
Ni3-N13	2.055(2)	N15-Ni4-N16	78.4(1)
Ni4-O3	2.144(2)	O8-N18-O9	120.7(3)
Ni4-O4	2.126(2)	O8-N18-O10	118.8(3)
Ni4-N11	1.987(2)	O9-N18-O10	120.5(3)
Ni4-N12	2.093(2)	O5-N17-O6	116.2(3)
Ni4-N15	1.979(2)	O5-N17-O7	122.2(4)
Ni4-N16	2.091(2)	O6-N17-O7	121.6(4)

Table 3. Selected bond lengths (Å) and angles for complex **4**.

Bond Lengths (Å) [a]		Bond Angles (°) [a]	
Co1-O1	1.899(2)	O1-Co1-N1	165.2(1)
Co1-N1	1.927(2)	O2-Co1-N5	165.4(1)
Co1-N2	1.855(2)	N2-Co1-N6	174.7(1)
Co1-O2	1.892(2)	O1-Co1-N2	82.5(1)
Co1-N5	1.915(2)	O1-Co1-O2	91.3(1)
Co1-N6	1.857(2)	O2-Co1-N6	82.6(1)
C8-O1	1.290(3)	N2-Co1-N5	100.7(1)
C8-N3	1.329(3)	O1-C8-N3	125.1(2)
N3-N2	1.391(3)	C8-N3-N2	106.3(2)
N2-C6	1.298(3)	N3-N2-C6	123.7(2)
C21-O2	1.306(3)	O2-C21-N7	124.2(2)
C21-N7	1.317(3)	C21-N7-N6	107.3(2)
N7-N6	1.383(3)	N7-N6-C19	123.9(2)
N6-C19	1.299(3)	O3-Cl1-O4	109.7(1)
Cl1-O3	1.440(2)	O3-Cl1-O6	110.7(1)
Cl1-O4	1.435(2)	O4-Cl1-O5	108.8(1)
Cl1-O5	1.442(2)	O5-Cl1-O6	108.8(1)
Cl1-O6	1.428(2)		

[a] The Cl-O bond lengths and O-Cl-O bond angles refer to the perchlorate counter ion not shown in Figure 5 and Figure S20.

Table 4. Selected interatomic distances (Å) and bond angles (°) for complex 3·0.8 H$_2$O·1.3MeOH.

Interatomic Distances (Å)		Bond Angles (°)	
Co1···Co2	3.467(2)	N1-Co1-N11	170.6(3)
Co1-N1	1.953(6)	N2-Co1-N5	172.5(3)
Co1-N2	1.906(6)	N6-Co1-N12	172.0(3)
Co1-N5	1.951(6)	N1-Co1-N2	81.3(3)
Co1-N6	1.899(6)	N2-Co1-N6	92.0(3)
Co1-N11	1.910(6)	N5-Co1-N6	81.6(3)
Co1-N12	1.931(6)	N11-Co1-N12	82.1(3)
Co2-N3	1.900(6)	N3-Co2-N9	170.3(3)
Co2-N4	1.935(6)	N4-Co2-N7	171.1(2)
Co2-N7	1.901(6)	N8-Co2-N10	171.4(2)
Co2-N8	1.946(6)	N3-Co2-N4	81.7(2)
Co2-N9	1.942(7)	N4-Co2-N8	93.1(2)
Co2-N10	1.909(6)	N7-Co2-N8	81.6(3)
C6-O1	1.242(9)	N8-Co2-N9	94.1(3)
C6-N2	1.363(9)	O1-C6-N2	126.9(7)
N2-N3	1.407(7)	C6-N2-N3	114.6(6)
N3-C7	1.310(8)	N2-N3-C7	121.6(6)
C19-O2	1.319(9)	O2-C19-N6	126.1(8)
C19-N6	1.311(8)	C19-N6-N7	119.4(6)
N6-N7	1.384(8)	N6-N7-C20	120.5(6)
N7-C20	1.331(9)	O3-C32-N10	124.9(9)
C32-O3	1.253(9)	C32-N10-N11	117.9(6)
C32-N10	1.348(8)	N10-N11-C33	121.3(6)
N10-N11	1.400(8)		
N11-C33	1.292(8)		

Two 2.21011 (Harris notation [35]) anionic L$^-$ ligands bridge adjacent Cu^{II} atoms on the short sides of the rectangle through their alkoxide oxygen atoms (O2, O2′). Two 2.11111 L$^-$ ligands bridge adjacent Cu^{II} atoms on the long sides of the rectangle with their diazine groups (N2-N3, N2′-N3′). The two different coordination modes are shown in Scheme 2. The Cu2/Cu2′ ions are 6-coordinate with a distorted octahedral geometry, the *trans* coordination angles being in the range 148.2(2)–175.5(2)°. The Jahn–Teller axis is defined as N8-Cu2-O2 [Cu2-N8 = 2.193(5) Å, Cu2-O2 = 2.331(4) Å]. The Cu1/Cu1′ ions are 5-coordinate with a {Cu^{II}N$_3$OBr} chromophore; the access to the sixth donor site is blocked by the presence of the methyl groups. The coordination geometry can be described as either distorted square pyramidal or distorted trigonal bipyramidal. Analysis of the shape-determining angles using the approach of Addison and Reedjik [36] yields a value for the trigonality index, τ, of 0.42 ($\tau = 0$ and 1 for square pyramidal and trigonal bipyramidal geometry, respectively). Adopting the square pyramidal description, the basal plane consists of donor atoms N2, N5, O2 and Br1, with atom N1 occupying the apical position. The alternative distorted trigonal bipyramidal description places donor atoms N2 and N5 at the axial positions, and atoms N1, O2 and Br1 at the equatorial sites. The Cu1-O2-Cu2 (Cu1′-O2′-Cu2′) angle is 138.6(2)°. Establishing the site of deprotonation is sometimes difficult for L$^-$ and related ligands. The C6-O1, C6-N2, N2-N3, N3-C8 and C19-O2, C19-N6, N6-N7, N7-C21 bond distances (Table 1) indicate a charge delocalization within the OCNNC backbone of the two crystallographically independent, deprotonated ligands of the tetranuclear cation. If we should define the deprotonated atom, the relatively long C6-O1 and C19-O2 distances [1.257(7) and 1.287(7) Å, respectively] suggest that the O atoms are the principal sites of deprotonation [14].

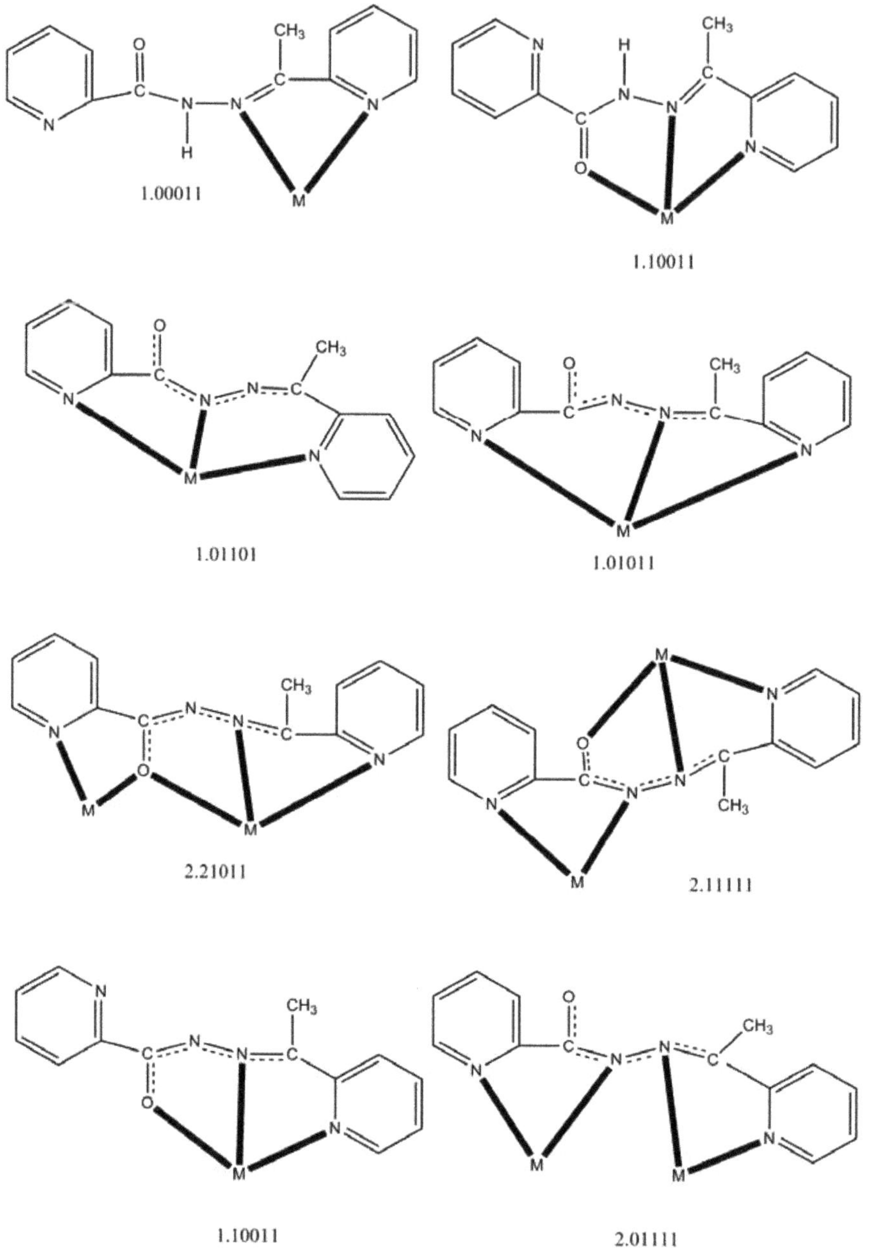

Scheme 2. The to date crystallographically established coordination modes of LH and L$^-$, and the Harris notation that describes these modes. The neutral ligand (LH) is shown in the keto-amino form. The central OCNNC unit of the anionic ligand has been drawn in a manner that emphasizes its delocalized description that appears in most complexes.

The tetranuclear cations of the complex form layers parallel to the (001) plane through π–π stacking interactions (Figure 2). The Cg1$\cdots$Cg1″ and Cg2$\cdots$Cg2‴ distances are 3.713(1) and 3.661(1) Å, respectively, where Cg1 is the centroid of the N1-containing aromatic ring and Cg2 is the centroid of the N4-containing ring [symmetry codes: ('') = $-x + 2, -y + 2, -z$; (''') = $-x + 1, -y + 1, -z$]. The cations that form the layers interact further through $\pi-\pi$ overlaps between centrosymmetrically-related ligands involving the N5, N8- and N5*, N8*-containing rings which are at a 3.34 Å distance and belong to adjacent layers (Figure S2), thus building the 3D architecture of the structure [symmetry code: (*) = $-x + 1, -y + 2, -z + 1$]. The Br$^-$ counterions, and the solvate H_2O and MeOH molecules are hosted in the lattice through H bonds (Table S18).

The crystal structure of **2**·0.2H_2O·3EtOH consists of tetranuclear cations $[Ni_4(NO_3)_2(L)_4(H_2O)]^{2+}$ (Figure 3 and Figure S19), NO_3^- counterions, as well as EtOH and H_2O molecules. The cation is a square [2 × 2] grid. The four metal ions are practically coplanar, the distances from their best mean plane being: Ni1 0.0016(4) Å, Ni2 0.0016(4) Å, Ni3 0.016(4) Å and Ni4 0.017(4) Å. The sides of the square are in the 3.895(1)–3.938(1) Å range, and the diagonals are 5.526(1) Å (Ni1···Ni4) and 5.559(1) Å (Ni2···Ni3).

The two Ni^{II} atoms of each side of the square are bridged by the alkoxide O atom (O1, O2, O3, O4) of one 2.21011 L$^-$ ligand (Scheme 2) and the core of the cluster cation is thus $\{Ni_4(OR)_4\}^{4+}$. One terminal H_2O ligand and one monodentate nitrato group are coordinated to Ni3, while the two vacant coordination sites at Ni2 are occupied by two oxygen atoms (O5, O6) of a bidentate chelating nitrato group. The metal ions are all six-coordinate with distorted octahedral geometries, and the chromophores are $\{Ni1N_4O_2\}$, $\{Ni2N_2O_4\}$, $\{Ni3N_2O_4\}$ and $\{Ni4N_4O_2\}$. The Ni-O and Ni-N bond lengths are typical [7,37,38] for octahedral Ni(II) complexes. The C-O$_{bridging}$ bond distances are in the narrow 1.295(3)–1.306(3) Å, indicating the predominance of single CO bond character [15]. The Ni-O-Ni angles are ~138°. Within the cation, there is a strong H bond with O1W as donor and the non-coordinate O9 atom of the nitrato ligand as acceptor (Figure 3, Table S3).

The tetranuclear cations of the complex form layers parallel to the (100) plane through $\pi-\pi$ stacking interactions and H bonds (Figure 4). The $\pi-\pi$ overlap of the N8- and N12'-containing rings of neighboring cluster cations create chains parallel to the b axis [symmetry code: (') = $-x, y - 1, z$]; the distance between their centroids is 4.076(1) Å and the angle between the ring planes is 9.2(2)°. Through the $\pi-\pi$ overlap of centrosymmetrically-related rings that contain N1 and N1'' [symmetry code: ('') = $-x + 2, -y - 1, -z - 1$; the distance between the planes is 3.36(2) Å] and belong to neighboring cluster cations, pairs of chains are formed extending parallel to the b axis and thus double chains are created. Cations belonging to neighboring double chains interact through C20-H$_C$(C20)···O10 and C15-H(C15)···O1W H bonds (Table S3), where C20 and C15 are methyl and aromatic carbon atoms, respectively, forming layers of double chains parallel to the (100) plane. These layers are stacked along the a axis. Nitrate counterions and lattice EtOH molecules are hosted between the layers through an extensive H-bonding network (Table S3); these species are linked with each other and also connect neighboring layers building the 3D architecture of the structure. The donors of these H bonds are mainly the oxygen atoms of the lattice H_2O and EtOH molecules, while the acceptors are counter nitrate and lattice EtOH oxygen atoms.

We start the structural descriptions of the Co(III) complexes with the simplest compound, i.e., **4**. The crystal structure of **4** consists of ions $[Co(L)_2]^+$ (Figure 5 and Figure S20) and ClO_4^- in an 1:1 ratio. Both mutually perpendicular L$^-$ ions act as tridentate chelating, meridional 1.10011 ligands (Scheme 2), each ligand forming two practically planar 5-membered chelating rings. The metal ion is coordinated by two alkoxide oxygen atoms (O1, O2), two 2-pyridyl nitrogen atoms (N1, N5) and two hydrazone nitrogen atoms (N2, N6) resulting in a distorted octahedral geometry. The *trans* angles of the octahedron are 165.2(1), 165.4(1) and 174.7(1)°. Due to the meridional character of each anionic ligand, two pairs of *trans* coordination sites are each occupied by donor atoms of the same L$^-$ group (O1/N1, O2/N5), whereas the third pair of coordination sites consists of nitrogen atoms (N2, N6) from different ligands. The Co-(O, N) bond lengths are in the range 1.855(2)–1.927(2) Å and agree very well with values observed for low-spin Co(III) ions in octahedral environments [35,39,40]. The two Co-N bond lengths for each ligand differ [1.927(2) vs. 1.855(2) Å, and 1.915(2) vs. 1.857(2) Å]; the shorter bond distance (stronger bond) pertains to the hydrazone nitrogen atoms (N2, N6), probably due to the presence of some negative charge on these atoms (because of delocalization).

The mononuclear cations of **4** form chains parallel to the [111] crystallographic direction through π–π stacking interactions between centrosymmetrically-related N(1)- and N(1′)-containing rings [symmetry code: (′) = $-x$, $-y + 1$, $-z + 1$; the distance between the planes is 3.62(1) Å] and between centrosymmetrically-related N(4)- and N(4′)-containing rings [symmetry code: (′) = $-x + 1$, $-y + 2$, $-z + 2$; the distance between the planes is 3.36(1) Å] (Figure 6). The cations of a given chain are further linked through C12-H(C12)$\cdots$N8 and C12-H(C12)$\cdots$O2 bifurcated H bonds (Table S4); C12 is an aromatic carbon atom. The chains are connected through C17-H(C17)$\cdots$N4, C1-H(C1)$\cdots$O5 and C13-H(C13)$\cdots$O5 H bonds forming layers parallel to the (1–10) plane; atoms C1, C13 and C17 are aromatic carbon atoms and O5 belongs to the ClO_4^- counterion. These layers are stacked along the b axis and linked through C14-H(C14)$\cdots$O4, C13-H(C13)$\cdots$O5 and C7-H$_B$(C7)$\cdots$N7 H bonds (Table S4); C14 is an aromatic carbon atom and C7 is a methyl carbon atom.

The crystal structure of **3**·0.8H$_2$O·1.3MeOH consists of dinuclear cations [Co$_2$(L)$_3$]$^{3+}$ (Figure 7 and Figure S21) and ClO_4^- counterions in an 1:3 ratio, as well as solvent H$_2$O and MeOH lattice molecules. The cation has approximate D$_3$ symmetry with the three bis(bidentate) 2.01111 L$^-$ ligands (Scheme 2) wrapped around the Co1$\cdots$Co2 axis in such a way as to give each metal ion a distorted octahedral {Co^{III}N$_6$} coordination. The Co^{III}-N bond lengths are in the range 1.899(6)–1.953(6) Å confirming [34,40] the low-spin character of the two 3d^6 metal ions. These bond distances follow the same trend seen for **4**, i.e., Co^{III}-N (hydrazone) < Co^{III}-N (2-pyridyl). The *trans* coordination angles for Co1 and Co2 are in the ranges 170.6(3)–172.5(3)° and 170.3(3)–171.4(2)°, respectively. Each six-coordinate Co^{III} atom adopts a *fac* configuration; the 2-pyridyl nitrogen atoms (and also the hydrazone nitrogen atoms) are arranged so as to define one face of the octahedron for each metal center. The Co1$\cdots$Co2 distance is rather short, i.e., 3.467(2) Å, due to the presence of three diatomic bridges between the two metal ions. The cation can be considered as a *pseudo* triple helicate. We prefer the term *pseudo* (or helicate-type) because ligands that form helicates generally comprise two bidentate coordinating groups A-B connected via some linker [41–46]; this linker is absent in L$^-$. In most cases (including our complex) the helicities of the two metal centers (i.e., Δ and Λ) may be mechanically coupled, so for example formation of a Δ configuration at the first metal ion induces Δ configuration at the second; this is clearly illustrated in Figure S21c for one of the [Co$_2$(L)$_3$]$^{3+}$ cations of compound **3**·0.8H$_2$O·1.3MeOH which is a homochiral Δ,Δ ("right-handed") system. Due to the centrosymmetric space group ($P\bar{1}$), both enantiomers (Δ,Δ and Λ,Λ) are present in the helical structure of the complex [41,43], i.e., the crystal is a racemic sample.

The dinuclear cations of **3**·0.8H$_2$O·1.3MeOH form chains parallel to the [101] crystallographic direction through C9-H(C9)$\cdots$O1 and C17-H(C17)$\cdots$O3 H bonds (Figure 8, Table S5). These chains build the 3D architecture of the complex through H bonds involving aromatic carbon atoms as donors, and ClO_4^- and lattice H$_2$O oxygen atoms as acceptors (Table S5).

Complexes **1–4** join a family of structurally characterized coordination complexes containing LH and L$^-$ as ligands. The to date characterized metal complexes are listed in Table 5 along with information about the coordination mode of the ligands, the nuclearity/dimensionality of the products and the coordination geometry of the metal ions involved. Perusal of Table 5 shows that **2**, **3** and **4** are the first structurally characterized nickel and cobalt complexes of any form (neutral or anionic) of the ligand. The L$^-$ coordination modes 2.01111 and 1.10011 (Scheme 2) have been crystallographically established for the first time in the Co(III) complexes **3** and **4**. The rectangular [2 × 2] grid complex [Cu$_4$Br$_2$(L)$_4$]Br$_2$ (**1**) is structurally similar to complex [Cu$_4$(L)$_4$(H$_2$O)$_2$](NO$_3$)$_4$ [14], albeit with slight differences resulting from the different nature of the ancillary inorganic ligand (Br$^-$ vs. H$_2$O). Many (but not all) structural characteristics of the square [2 × 2] grid complex [Ni$_4$(NO$_3$)$_2$(L)$_4$H$_2$O](NO$_3$)$_2$ (**2**) resemble those of the Mn(II) complexes [Mn$_4$(CF$_3$SO$_3$)(L)$_4$(H$_2$O)$_3$](CF$_3$SO$_3$)$_3$ [15] and [Mn$_4$(N$_3$)$_4$(L)$_4$] [19].

Table 5. To date crystallographically characterized metal complexes of LH and L$^-$, and relevant structural information.

Complex [a]	Coordination Mode [b,c]	Nuclearity/Dimensionality	Coordination Geometry	Ref.
[CdBr$_2$(LH)]	η^1:η^1:η^1(1.10011)	Mononuclear	Trigonal bipyramidal	[16]
[Ln(NO$_3$)$_3$(LH)(MeOH)$_2$] (Ln = La, Ce)	η^1:η^1:η^1(1.10011)	Mononuclear	Capped pentagonal antiprismatic	[17]
[Nd(NO$_3$)$_3$(LH)(H$_2$O)]	η^1:η^1:η^1(1.10011)	Mononuclear	Bicapped square antiprismatic	[17]
[PdCl$_2$(LH)]	η^1:η^1(1.01100)	Mononuclear	Square planar	[18]
[HgX$_2$(LH)] (X = Cl, Br)	η^1:η^1:η^1(1.10011)	Mononuclear	Square pyramidal	[20]
[HgI$_2$(LH)(H$_2$O)]	η^1:η^1:η^1(1.10011)	Mononuclear	Octahedral	[20]
{[Pb$_3$Br$_6$(LH)$_2$]}$_n$	η^1:η^1:η^1(1.10011)	1D metal-organic ribbon	7-coordinate [d], Octahedral	[21]
[PdCl(L)]	η^1:η^1:η^1(1.01011)	Mononuclear	Square planar	[18]
[PdCl(L)]	η^1:η^1:η^1(1.01101)	Mononuclear	Square planar	[18]
[Cu$_4$(L)$_4$(H$_2$O)$_2$](NO$_3$)$_4$	η^1:η^2:η^1:η^1:μ_2(2.21011), η^1:η^1:η^1:η^1:μ_2(2.11111)	Rectangular [2 × 2] grid	Square pyramidal, Octahedral	[14]
[Mn$_4$(CF$_3$SO$_3$)(L)$_4$(H$_2$O)$_3$](CF$_3$SO$_3$)$_3$	η^1:η^2:η^1:η^1:μ_2(2.21011)	Square [2 × 2] grid	Octahedral	[15]
[Mn$_5$(L)$_6$](ClO$_4$)$_4$	η^1:η^2:η^1:η^1:μ_2(2.21011)	Trigonal bipyramidal {Mn$_5$(μ_2-OR)$_6$}$^{4+}$ core	Octahedral	[15]
[Mn$_4$(N$_3$)$_4$(L)$_4$]	η^1:η^2:η^1:η^1:μ_2(2.21011)	Square [2 × 2] grid	Octahedral	[19]
[Cu$_4$Br$_2$(L)$_4$]Br$_2$ (1)	η^1:η^2:η^1:η^1:μ_2(2.21011), η^1:η^1:η^1:η^1:μ_2(2.11111)	Rectangular [2 × 2] grid	Square pyramidal, Octahedral	this work
[Ni$_4$(NO$_3$)$_2$(L)$_4$(H$_2$O)](NO$_3$)$_2$ (2)	η^1:η^2:η^1:η^1:μ_2(2.21011)	Square [2 × 2] grid	Octahedral	this work
[Co$_2$(L)$_3$](ClO$_4$)$_3$ (3)	η^1:η^1:η^1:η^1:μ_2(2.01111)	Dinuclear helicate	Octahedral	this work
[Co(L)$_2$](ClO$_4$) (4)	η^1:η^1:η^1(1.10011)	Mononuclear	Octahedral	this work

[a] Lattice solvent molecules have been omitted. [b] Both the η/μ convention and the Harris notation are provided. [c] See Scheme 2. [d] The coordination geometry of the 7-coordinate Pb^{II} Atom was not mentioned.

2.4. Magnetochemistry

Direct current (dc) magnetic susceptibility data (χ) on dried polycrystalline, analytically pure samples of **1** and **2** were collected in the 1.8–310 K range in an applied field of 1 kOe. The data are plotted as χ vs. T and χT vs. T plots in Figure 9; Figure 10. Magnetization data were collected as a function of field from 0 to 50 kOe at 1.8 K and are presented in Figures S22 and S23. Although M vs. H per mole may be presented in units of Bohr magneton, we present the data in units of emu/mol because current literature uses these units commonly.

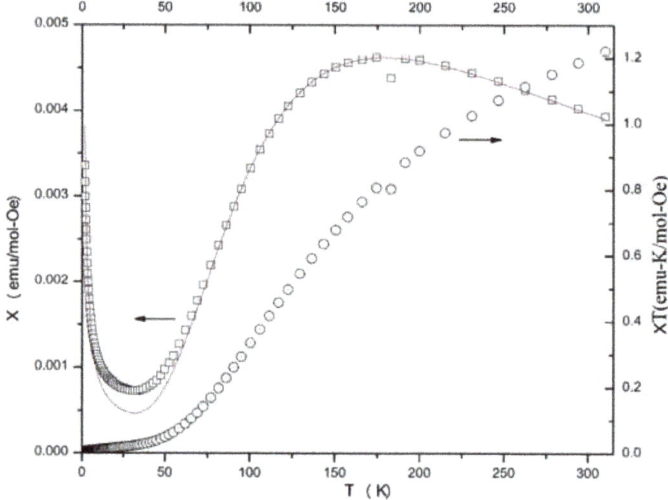

Figure 9. χ vs. T and χT vs. T plots for compound **1**. Arrows on the plots indicate which y-axis. Applies to which data. The solid line is the best fit of the χ vs. T data to the $S = 1/2$ dimer model; see the text for details.

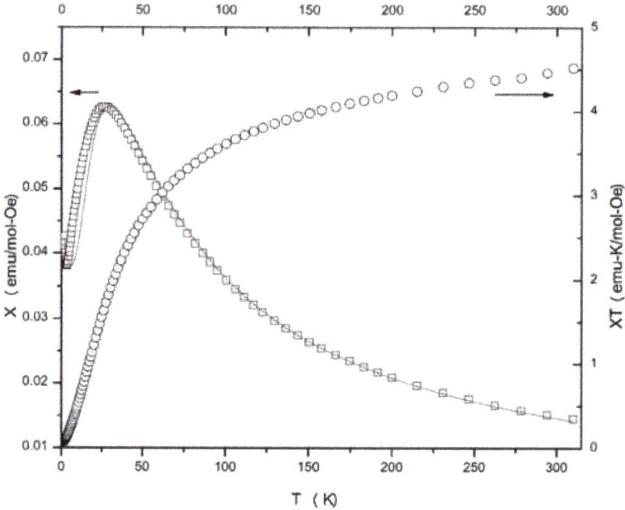

Figure 10. χ vs. T and χT vs. T plots for compound **2**. Arrows on the plots indicate which y-axis applies to. Which data. The solid line is the best fit of the χ vs. T data to the cyclic $S = 1$ tetramer model; see the text for details.

Data for **1** show a maximum in χ near 175 K indicative of significant antiferromagnetic interaction (Figure 9). The χT decreases as T decreases reaching ~0 emu K/mol Oe at ~25 K. Magnetization for this complex reaches a saturation value of 36 emu/mol at 30 kOe, which is clearly too small for the bulk sample (Figure S22). A saturation value of nearly 24,000 emu/mol for four Cu^{II} ions would be expected, indicating that the bulk sample is in a singlet ground state at 1.8 K and only a minor paramagnetic impurity is providing the observed moment. Susceptibility data for **1** were fit to a model for an $S = 1/2$ dimer using MAGMUN 4.1 [47]. The quality of fit is high through the maximum in χ, but begins to deviate below 75 K, where the magnetism is dominated by the contributions of a trace magnetic impurity. The results yield a Curie constant (C) of 1.66(3) emu K/mol Oe ($0.41/Cu^{II}$), $J = -198(2)$ K (using the Hamiltonian $H = -J\Sigma S_1 \cdot S_2$) and $\rho = 0.4(1)\%$ (ρ is the paramagnetic impurity). Attempts to fit the data to the same model with a Curie–Weiss correction to account for interdimer interactions yielded virtually identical results and a ϑ value of $-0.9(2)$ indicating that the interdimer interactions are not significant.

The symmetry and rectangular nature of the complex (Figure 1 and Figure S17) suggest a model with two J values, one for the diazine (or hydrazonate)-mediated exchange and one for the alkoxido-mediated exchange. Attempts to fit the data to such a model with two J values resulted in $J_1 = -195(2)$ K and $J = -1(1)$ K, again indicating that the exchange *between* the diazine-bridged dimeric units of the tetranuclear cation is negligible. This observation is in very good agreement with results for the structurally similar complex $[Cu_4(L)_4(H_2O)_2](NO_3)_4$ by the group of Thompson [14]. As in the case of the present complex **1**, the Cu^{II} centers in the nitrate complex exist as diazine-bridged pairs of alkoxido-bridged dimers and they reported a value of ~ -215 K through the diazine bridge and 0 K through the oxygen bridge. The $J_2 = ~ 0$ K value is consistent with the orthogonal alkoxide bridging arrangement [14]. The J_1 value for **1** is entirely consistent with the large (155.3°) Cu-N-N-Cu torsion angle at the diazine bridge and the well established correlations involving this angle and exchange integral for a series of dinuclear copper(II) complexes [14,48,49] in which the metal ions are bridged by a diazine group. Thus, **1** can be considered as a practically isolated pair of antiferromagnetically coupled dinuclear units.

The magnetization of compound **2** reaches a value of ~1400 emu/mol at 50 kOe, well below the expected saturation value for four Ni^{II} ions (~48,000 emu/mol), but the magnetization is still clearly rising (Figure S23) indicating that a significantly larger field would be needed to saturate the sample. This suggests the presence of measurable antiferromagnetic exchange in the complex, but also shows that the bulk material is still paramagnetic at 1.8 K, unlike sample **1**. The χT product decreases as T decreases rather smoothly in the 310–100 K range, and then more rapidly in the 100–1.8 K range reaching a value of ~0.05 emu K/mol Oe at 1.8 K.

Susceptibility data for **2** were fit to a model for an $S = 1$ cyclic (square) tetramer using MAGMUN 4.1 [47]. The fit is qualitatively acceptable, but overestimates the temperature of the maximum in χ by ~5 K, although the value of χ at χ_{max} is well reproduced (Figure 10). The fitted values are Curie constant = 4.61(3) emu K/mol Oe ($1.15/Ni^{II}$), $J = -12.6(1)$ K ($H = -J\Sigma S_1 \cdot S_2$), $\rho = 0.25(5)$ % and $D = 22.5(5)$ K. Attempts to fit the data to a simple $S = 1$ dimer model were unsuccessful. Although somewhat large, the fitted value for the single-ion anisotropy of the Ni^{II} ions ($D = 22.5(5)$ K) is not without precedent [50–52] and substantially larger values have been reported [53–55]. It is clear that the structure has lost its solvate lattice molecules, based on the analytical data and on the powder X-ray diffraction pattern for the sample used for magnetic data collection, and as a result the symmetry of the system could have been reduced. The quality of fit suggests that the cyclic $S = 1$ model is reasonable, but the slight change in structure may render the superexchange pathways inequivalent and thus the values presented likely represent an average of J, C and D values. The inequivalence of the superexchange pathways might also be due to the different nature of the terminal donor atoms for the Ni^{II} centers of the cation $[Ni_4(NO_3)_2(L)_4(H_2O)]^{2+}$. In any case, and especially in the absence of a better understanding of the structure of the desolvated material, additional detailed analysis would likely lead to an overinterpretation of the data available.

The antiferromagnetic J coupling between the Ni^{II} centers in the cation of cluster **2** is clearly associated with the large Ni-O-Ni angles [7,39,56,57]; the Ni1-O1-Ni2, Ni2-O3-Ni4, Ni4-O4-Ni3 and Ni3-O2-Ni1 (Figure 3 and Figure S3) are 138.4(1), 138.2(1), 137.5(1) and 138.5(1)°, respectively.

3. Experimental Section

3.1. Materials, Physical and Spectroscopic Measurements

All manipulations were performed under aerobic conditions using reagents and solvents (Alfa Aesar, Karlsruhe, Germany; Aldrich, Tanfrichen, Germany) as received. The organic ligand LH was synthesized in typical yields of >95% as described in the literature [14–16,18], i.e., by the 1:1 reaction between picolinic acid hydrazide and 2-acetylpyridine in refluxing EtOH for 3 h. Its purity was checked by microanalyses (C, H, N), determination of the melting point (found, 193–194 °C; reported, 195–197 °C), and ^{1}H NMR and IR spectra. Elemental analyses (C, H, N) were performed by the University of Patras (Patras, Greece) microanalytical service. Fourier transform infrared (FT–IR) spectra were recorded using a Perkin-Elmer 16PC spectrometer (Perkin-Elmer, Watham, MA, USA) with samples prepared as KBr pellets and as nujol or hexochlorobutadiene mulls between CsI disks. ^{1}H NMR spectra of the diamagnetic Co(III) complexes were recorded on a 400 MHz Bruker Avance DPX spectrometer (Bruker, Karlsruhe, Germany) using $(Me)_4Si$ as internal standard. UV/VIS solution spectra were recorded using a Specord 50 Plus spectrophotometer (Analytik Jena, Jena, Germany). Magnetic susceptibility data were collected using a Quantum Design MPMS-XL SQUID magnetometer (San Diego, CA, USA). Samples of **1** and **2** were ground and loaded into gelatin capsules. Magnetization data were collected as a function of field from 0 to 50 kOe at 1.8 K. Several data points were collected as the field was reduced back to zero to check for hysteresis effects; none were observed. Susceptibility data were collected for the background signal of the sample holder (measured independently), for the diamagnetic contributions of the constituent atoms as estimated via Pascal's constants [58], and for the temperature-independent paramagnetism of the Cu^{II} and Ni^{II} ions.

3.2. Synthesis of Complex $[Cu_4Br_2(L)_4]Br_2 \cdot 0.8H_2O \cdot MeOH$ ($1 \cdot 0.8H_2O \cdot MeOH$)

To a stirred solution of LH (0.048 g, 0.20 mmol) in MeOH (20 mL) were added solids NaO_2CPh (0.029 g, 0.20 mmol) and $CuBr_2$ (0.045 g, 0.20 mmol). The resulting green slurry was stirred at room temperature for a further 30 min, filtered to remove an amount of NaBr and the green-brown filtrate was left undisturbed in a closed flask. X-ray quality, greenish brown crystals of the product were formed over a period of 3 days. The crystals were collected by filtration, washed with cold MeOH (2 × 1 mL) and Et_2O (3 × 2 mL), and dried in a vacuum dessicator over P_4O_{10} overnight. The yield was 62%. The complex was satisfactorily analyzed as lattice solvent-free, i.e., as **1**. Analyses calculated for $C_{52}H_{44}N_{16}O_4Cu_4Br_4$ (found values in parentheses): C 40.79 (41.02), H 2.90 (2.84), N 14.64 (14.50) %. IR bands (KBr, cm^{-1}): 3080w, 3035w, 2960w, 1594m, 1576sh, 1534s, 1474s, 1436w, 1370s, 1322w, 1290m, 1258m, 1180m, 1144w, 1100w, 1080w, 1042m, 1016m, 920m, 804sh, 780m, 752m, 716m, 710m, 688m, 654sh, 640w, 574w, 565w, 504w, 462w. UV/VIS bands (MeOH, nm): 245, 280sh, 370, ~745.

3.3. Synthesis of Complex $[Ni_4(NO_3)_2(L)_4(H_2O)](NO_3)_2 \cdot 0.2H_2O \cdot 3EtOH$ ($2 \cdot 0.2H_2O \cdot 3EtOH$)

To a stirred slurry of LH (0.048 g, 0.20 mmol) in CH_2Cl_2 (2 mL) was added a green solution of $Ni(NO_3)_2 \cdot 6H_2O$ (0.058 g, 0.20 mmol) in EtOH (10 mL). The resulting greenish brown solution was stirred at room temperature for a further 45 min, filtered and the filtrate was allowed to stand undisturbed in a closed flask. X-ray quality brown crystals of the product were precipitated over a period of two weeks. The crystals were collected by filtration, washed with cold EtOH (2 × 2 mL) and Et_2O (5 × 3 mL), and dried in a vacuum dessicator over anhydrous $CaCl_2$. Typical yields were in the range 30–35%. The complex was satisfactorily analyzed as lattice solvent-free, i.e., as **2**. Analyses calculated for $C_{52}H_{46}N_{20}O_{17}Ni_4$ (found values in parentheses): C 42.84 (42.67), H 3.19 (3.26), N 19.22 (18.87) %. IR bands (KBr, cm^{-1}): 3420mb, 3070w, 3030w, 2950w, 1598w, 1560w, 1522m, 1466m, 1438w,

1384s, 1368sh, 1294w, 1260w, 1182w, 1162w, 1144w, 1102w, 1066w, 1048w, 1026w, 922w, 810w, 780w, 761w, 714w, 694w, 642w, 562w, 505w, 420w. UV/VIS bands (MeOH, nm): 250sh, 290sh, 365, 615, ~980. The same complex can be prepared -in comparable yields- using 1 equiv. of Et_3N per ligand in the reaction mixture.

3.4. Synthesis of Complex $[Co_2(L)_3](ClO_4)_3 \cdot 0.8H_2O \cdot 1.3MeOH$ ($3 \cdot 0.8H_2O \cdot 1.3MeOH$)

To a stirred yellow solution of LH (0.024 g, 0.10 mmol) in MeOH (10 mL) was added solid $Co(ClO_4)_2 \cdot 6H_2O$ (0.037 g, 0.10 mmol). The solid soon dissolved and the resulting brown solution was stirred overnight at room temperature, filtered and the filtrate was allowed to stand undisturbed in a closed flask. X-ray quality brown crystals of the product were precipitated over a period of 3–4 days. The crystals were collected by filtration, washed with cold MeOH (1 mL) and Et_2O (3 × 1 mL), and dried in air. The yield was 48% (based on the available LH). The complex was satisfactorily analyzed as $[Co_2(L)_3] \cdot H_2O$, i.e., as $3 \cdot H_2O$. Analyses calculated for $C_{39}H_{35}N_{12}O_{16}CoCl_3$ (found values in parentheses): C 40.66 (40.54), H 3.07 (3.12), N 14.59 (14.74) %. IR bands (KBr, cm^{-1}): 3422mb, 3094w, 2951w, 2902w, 1606m, 1508m, 1458m, 1436m, 1376m, 1334w, 1300w, 1258w, 1178m, 1105sh, 1088s, 918w, 806w, 765sh, 756w, 718w, 688w, 662w, 624m, 510w, 412w. UV/VIS bands (MeCN, nm): 295, 395, 440, 580, 735. ^1H NMR peaks (DMSO-d_6, δ/ppm): 8.59(d, 3H), 8.25(dd, 6H), 8.18(d, 3H), 8.02(d, 3H), 7.93(t, 3H), 7.55 (mt, 6H), 3.35(s, 9H), 3.17(s, 3H). The same complex can be prepared-in slightly higher yields(~55%)-by the addition of 1 equiv. of Et_3N per ligand in the reaction mixture.

3.5. Syntheses of Complex $[Co(L)_2](ClO_4)$ (**4**)

$Co(ClO_4)_2 \cdot 6H_2O$ (0.037 g, 0.10 mmol) and LH (0.072 g, 0.30 mmol) were dissolved in MeOH (9 mL). To the resulting red solution, Et_3N (0.042 mL, 0.30 mmol) was slowly added. The solution became dark brown-red, was stirred overnight, filtered and left undisturbed in a closed flask. X-ray quality brown crystals of the product were precipitated over a period of 2–3 days. The crystals were collected by filtration, washed with cold MeOH (2 mL) and Et_2O (10 × 2 mL), and dried in air. Typical yields were in the 50–55% range (based on the available cobalt). Analyses calculated. for $C_{26}H_{22}N_8O_6CoCl$ (found values in parentheses): C 49.03 (49.37), H 3.49 (3.45), N 17.60 (16.99) %. IR bands (KBr, cm^{-1}): 3074w, 3012w, 2956w, 2933w, 1602m, 1496s, 1474sh, 1450s, 1430sh, 1372s, 1328m, 1310w, 1292w, 1256w, 1170m, 1118s, 1092sh, 1080s, 994w, 916w, 812w, 782m, 772sh, 754w, 742w, 718m, 704m, 662w, 624m, 512w, 495w, 429w. UV/VIS bands (MeCN, nm): 285, 405, 450, 590, ~760. ^1H NMR peaks (DMSO-d_6, δ/ppm): 8.58(d, 2H), 8.24(q, 4H), 8.15(d, 2H), 8.02(d, 2H), 7.88(t, 2H), 7.53 (mt, 4H), 3.34(s, 6H). Complex **4** can also be prepared by the 1:1 reaction between **3** and LH in refluxing MeOH in the absence of external base (yield <20%) and in the presence of LiOH (yield ~65%), Equations (6) and (7), respectively (*vide supra*).

3.6. Single-Crystal X-ray Crystallography

Suitable crystals of $1 \cdot 0.8H_2O \cdot MeOH$ (0.09 × 0.13 × 0.18 mm), $2 \cdot 0.2H_2O \cdot 3EtOH$ (0.22 × 0.32 × 0.55 mm), $3 \cdot 0.8H_2O \cdot 1.3MeOH$ (0.07 × 0.12 × 0.17 mm) and **4** (0.19 × 0.29 × 0.30 mm) were taken from the mother liquor and immediately cooled to −113 °C ($3 \cdot 0.8H_2O \cdot 1.3MeOH$) and −103 °C (for the other three compounds). Diffraction data were collected on a Rigaku R-AXIS SPIDER Image Plate diffractometer using graphite-monochromated Mo Kα ($1 \cdot 0.8H_2O \cdot MeOH$, $2 \cdot 0.2H_2O \cdot 3EtOH$) or Cu K$\alpha$ ($3 \cdot 0.8H_2O \cdot 1.3MeOH$, **4**) radiation. Data collection (ω-scans) and processing (cell refinement, data reduction and empirical absorption correction) were performed using the CrystalClear package [59]. The structures were solved by direct methods using SHELXS-97 [60] and refined by full-matrix least-squares techniques on F^2 with SHELXL, ver. 2014/6 [61]. Important crystallographic and refinement details are listed in Table S1. All non-H atoms were refined anisotropically. The H atoms of the four structures were either located by difference maps and refined isotropically or were introduced at calculated positions and refined as riding on their corresponding bonded atoms. Plots of the structures were drawn using the Diamond 3 program package [62].

The X-ray crystallographic data for the complexes have been deposited with CCDC (reference CCDC 1915158, 1915160, 1915159 and 1915157 for **1**·0.8H$_2$O·MeOH, **2**·0.2H$_2$O·3EtOH, **3**·0.8H$_2$O·1.3MeOH and **4**, respectively). They can be obtained free of charge at http://www.ccdc.cam.ac.uk/conts/retrieving.html or from the Cambridge Crystallographic Data Centre, 12 Union Road, Cambridge, CB2 1EZ, UK: Fax: +44-1223-336033; or e-mail: deposit@ccdc.cam.ac.uk.

4. Concluding Comments and Perspectives

It is rather difficult to conclude on a project that is still at its infancy. The present work extends the body of results (Table 5) that emphasize the ability of L$^-$ to form interesting structural types in 3d-metal chemistry. [2 × 2] rectangular (**1**) and square (**2**) grids have been characterized, while the interesting triple helicate-type dinuclear complex **3** was also isolated. Complexes **2–4** are the first, structurally characterized cobalt and nickel complexes of LH or L$^-$, while the two L$^-$ coordination modes in the Co(II) complexes (Table 5, Scheme 2) have been confirmed for the first time, emphasizing the flexibility and versatility of this ditopic ligand. The magnetic properties of **1** and **2** have been interpreted using one exchange interaction, and the former can be described as consisting of two antiferromagnetically coupled dinuclear units.

We believe that the research described herein has not exhausted new results. Indeed, studies in progress are producing additional products with other, magnetically interesting 3d-metal ions; our belief is that we have scratched only the surface of the coordination chemistry of LH/L$^-$. As far as future perspectives are concerned, we shall try to prepare lanthanide(III) clusters (only mononuclear complexes with the neutral ligand are known [17]; see also Table 5) and 3d/4f-metal complexes, based on L$^-$, with interesting magnetic properties. We are also trying to isolate complexes with ditopic ligands that are similar to LH, but with groups other than the methyl group (Scheme 1), because it is currently not evident whether the preparation and stability of 3d-metal complexes are dependent on the particular nature of the R substituent on the carbon atom next to the 2-pyridyl group.

Supplementary Materials: The following are available online at http://www.mdpi.com/: Figures S1–S4: IR spectra of the free ligand and representative complexes. Figures S5–S12: Solution UV/VIS/Near-IR electronic spectra of the complexes. Figures S13–S16: ^{1}H NMR spectra of the diamagnetic Co(III) complexes in DMSO-d$_6$. Figure S17: ORTEP plot of the tetranuclear cation [Cu$_4$Br$_2$(L)$_4$]$^{2+}$. Figure S18: The 3D arrangement of complex **1**·0.8H$_2$O·MeOH. Figure S19: ORTEP plot of the tetranuclear cation [Ni$_4$(NO$_3$)$_2$(L)$_4$(H$_2$O)]$^{2+}$. Figure S20: ORTEP plot of the mononuclear cation [Co(L)$_2$]$^+$. Figure S21: Various structural plots of the cation [Co$_2$(L)$_3$]$^{3+}$. Figure S22: Magnetization data for complex **1** at 1.8 K. Figure S23: Magnetization data for complex **2** at 1.8 K. Table S1: Crystallographic data for the four complexes. Tables S2–S5: H-bonding interactions in the crystal structures of the complexes.

Author Contributions: E.P., E.S. and E.-K.M. contributed toward the syntheses, crystallization, and conventional characterization (IR, UV/VIS, ^{1}H NMR) of the metal complexes. M.M.T. performed the magnetic measurements, interpreted the results and wrote the relevant part of the paper. C.P.R. and V.P. collected single-crystal X–ray crystallographic data, solved the structures and performed their refinement; the latter also studied in detail the supramolecular features of the crystal structures and wrote the relevant part of the article. S.P.P. coordinated the research, contributed to the interpretation of the results, and wrote the paper based on the reports of his collaborators. All the authors exchanged opinions concerning the interpretation and study of the results, and commented on the manuscript at all stages.

Funding: This research received no external funding.

Acknowledgments: Spyros P. Perlepes is grateful to the COST Action: CA15128-Molecular Spintronics (MOLSPIN) for encouraging his research activities in Patras.

Conflicts of Interest: The authors declare no conflict of interest.

References

1. Housecroft, C.E.; Sharpe, A.G. *Inorganic Chemistry*, 5th ed.; Pearson: Harlow, UK, 2018; pp. 237–239.
2. Ribas Gispert, J. *Coordination Chemistry*; Wiley-VCH: Weinheim, Germany, 2008; pp. XXXIX, XL.
3. Constable, E.C. *Metals and Ligand Reactivity*; VCH: Weinheim, Germany, 1996.

4. Stamatatos, T.C.; Efthymiou, C.G.; Stoumpos, C.C.; Perlepes, S.P. Adventures in the Coordination Chemistry of Di-2-pyridyl Ketone and Related Ligands: From High-Spin Molecules and Single-Molecule Magnets to Coordination Polymers, and from Structural Aesthetics to an Exciting New Reactivity Chemistry of Coordinated Ligands. *Eur. J. Inorg. Chem.* **2009**, 3361–3391. [CrossRef]

5. Kahn, O. *Molecular Magnetism*; VCH Publishers: New York, NY, USA, 1993.

6. Hearne, N.; Turnbull, M.M.; Landee, C.P.; van der Merwe, E.M.; Rademeyer, M. Halide-bi-bridged polymers of amide substituted-pyridines and –pyrazines: Polymorphism, structures, thermal stability and magnetism. *CrystEngComm* **2019**, *21*, 1910–1927. [CrossRef]

7. Efthymiou, C.G.; Mylonas-Margaritis, I.; Raptopoulou, C.P.; Psycharis, V.; Escuer, A.; Papantriantafyllopoulou, C.; Perlepes, S.P. A Ni_{11} Coordination Cluster from the Use of the Di-2-Pyridyl Ketone/Acetate Ligand Combination: Synthetic, Structural and Magnetic Studies. *Magnetochemistry* **2016**, *2*, 30. [CrossRef]

8. Escuer, A.; Esteban, J.; Perlepes, S.P.; Stamatatos, T.C. The bridging azido ligand as a central "player" in high-nuclearity 3d-metal cluster chemistry. *Coord. Chem. Rev.* **2014**, *275*, 87–129. [CrossRef]

9. Lada, Z.G.; Katsoulakou, E.; Perlepes, S.P. Synthesis and Chemistry of Single-molecule Magnets. In *Single-Molecule Magnets: Molecular Architectures and Building Blocks for Spintronics*; Holynska, M., Ed.; Wiley-VCH: Weinheim, Germany, 2019; pp. 245–313.

10. Milway, V.A.; Niel, V.; Abedin, T.S.M.; Xu, Z.; Thompson, L.K.; Grove, H.; Miller, D.O.; Parsons, S.R. Octanuclear and Nonanuclear Supramolecular Copper(II) Complexes with Linear "Tritopic" Ligands: Structural and Magnetic Studies. *Inorg. Chem.* **2004**, *43*, 1874–1884. [CrossRef]

11. Zhao, L.; Xu, Z.; Grove, H.; Milway, V.A.; Dawe, L.N.; Abedin, T.S.M.; Thompson, L.K.; Kelly, T.L.; Harvey, R.G.; Miller, D.O.; et al. Supramolecular Mn(II) and Mn(II)/Mn(III) Grid Complexes with $[Mn_9(\mu_2\text{-}O)_{12}]$ Core Structures. Structural, Magnetic, and Redox Properties and Surface Studies. *Inorg. Chem.* **2004**, *43*, 3812–3824. [CrossRef]

12. Milway, V.A.; Abedin, S.M.T.; Thompson, L.K.; Miller, D.O. Ferromagnetic Cu_8 pinwheel structures as building blocks for larger magnetic networks: Long-range structural and magnetic ordering. *Inorg. Chim. Acta* **2006**, *359*, 2700–2711. [CrossRef]

13. Milway, V.A.; Abedin, S.M.T.; Niel, V.; Kelly, T.L.; Dawe, L.N.; Dey, S.K.; Thompson, D.W.; Miller, D.O.; Alam, M.S.; Müller, P.; et al. Supramolecular 'flat' Mn_9 grid complexes—Towards functional molecular platforms. *Dalton Trans.* **2006**, 2835–2851. [CrossRef]

14. Grove, H.; Kelly, T.L.; Thompson, L.K.; Zhao, L.; Xu, Z.; Abedin, T.S.M.; Miller, D.O.; Goeta, A.E.; Wilson, C.; Howard, J.A.K. Copper (II) Complexes of a Series of Alkoxy Diazine Ligands: Mononuclear, Dinuclear and Tetranuclear Examples with Structural, Magnetic and DFT Studies. *Inorg. Chem.* **2004**, *43*, 4278–4288. [CrossRef]

15. Dawe, L.N.; Abedin, T.S.M.; Kelly, T.K.; Thompson, L.K.; Miller, D.O.; Zhao, L.; Wilson, C.; Leech, M.A.; Howard, J.A.K. Self-assembled polymetallic square grids ([2×2] M_4, [3×3] M_9) and trigonal bipyramidal clusters (M_5)-structural and magnetic properties. *J. Mater. Chem.* **2006**, *16*, 2645–2659. [CrossRef]

16. Akkurt, M.; Khandar, A.A.; Tahir, M.N.; Yazdi, S.A.H.; Afkhami, F.A. Dibromido {N′-[1-pyridin-2-yl) ethylidene]picolinohydrazide-$\kappa^2 N',O$} cadmium. *Acta Crystallogr. E* **2012**, *68*, m842.

17. Xu, Z.-Q.; Mao, X.-J.; Zhang, X.; Cai, H.-X.; Bie, H.-Y.; Xu, J.; Jia, L. Syntheses, Crystal Structures and Antitumor Activities of Three Ln(III) Complexes with 2-Acetylpyridine Picolinohydrazone. *Chin. J. Inorg. Chem.* **2015**, *31*, 1–8.

18. Kitamura, F.; Sawaguchi, K.; Mori, A.; Takagi, S.; Suzuki, T.; Kobayashi, A.; Kato, M.; Nakajima, K. Coordination Structure Conversion of Hydrazone-Palladium(II) Complexes in the Solid State and in Solution. *Inorg. Chem.* **2015**, *54*, 8436–8448. [CrossRef]

19. Abedi, M.; Yesilel, O.Z.; Mahmoudi, G.; Bauzá, A.; Lofland, S.E.; Yerli, Y.; Kaminsky, W.; Garczarek, P.; Zareba, I.K.; Ienco, A.; et al. Tetranuclear Manganese(II) Complexes of Hydrazone and Carboxydrazone Ligands: Synthesis, Crystal Structures, Magnetic Properties, Hirshfeld Surface Analysis and DFT Calculations. *Inorg. Chim. Acta* **2016**, *443*, 101–109. [CrossRef]

20. Mahmoudi, G.; Bauzá, A.; Gurbanov, A.V.; Zubkov, F.I.; Maniukiewicz, W.; Rodriguez-Dieguez, A.; López-Torres, E.; Frontera, A. The role of unconventional stacking interactions in the supramolecular assemblies of Hg(II) coordination compounds. *CrystEngComm* **2016**, *18*, 9056–9066. [CrossRef]

21. Mahmoudi, G.; Stilinovic, V.; Bauzá, A.; Frontera, A.; Bartyzel, A.; Ruiz-Pérez, C.; Kirillov, A.M. Inorganic-organic hybrid materials based on $PbBr_2$ and pyridine-hydrazone blocks—Structural and theoretical study. *RSC Adv.* **2016**, *6*, 60385–60393. [CrossRef]

22. Maniaki, D.; Pilichos, E.; Perlepes, S.P. Coordination Clusters of 3d-metals That Behave as Single-Molecule Magnets (SMMs): Synthetic Routes and Strategies. *Front. Chem.* **2018**, *6*, 461. [CrossRef]

23. Pilichos, E.; Mylonas-Margaritis, I.; Kontos, A.P.; Psycharis, V.; Klouras, N.; Raptopoulou, C.P.; Perlepes, S.P. Coordination and metal ion-mediated transformation of a polydentate ligand containing oxime, hydrazone and picolinoyl functionalities. *Inorg. Chem. Commun.* **2018**, *94*, 48–52. [CrossRef]

24. Lada, Z.G.; Sanakis, Y.; Raptopoulou, C.P.; Psycharis, V.; Perlepes, S.P.; Mitrikas, G. Probing the electronic structure of a copper(II) complex by CW- and pure—EPR spectroscopy. *Dalton Trans.* **2017**, *46*, 8458–8475. [CrossRef]

25. Anastasiadis, N.C.; Granadeiro, C.M.; Klouras, N.; Cunha-Silva, L.; Raptopoulou, C.P.; Psycharis, V.; Bekiari, V.; Balula, S.S.; Escuer, A.; Perlepes, S.P. Dinuclear Lanthanide(III) Complexes by Metal-Ion-Assisted Hydration of Di-2-pyridyl Ketone Azine. *Inorg. Chem.* **2013**, *52*, 4145–4147. [CrossRef]

26. Duros, V.; Sartzi, H.; Teat, S.J.; Sanakis, Y.; Roubeau, O.; Perlepes, S.P. Tris{2,4-bis(2-pyridyl)-1,3,5-triazapentanedionato}manganese(III), a complex derived from a unique metal ion-assisted transformation of pyridine-2-amidoxime. *Inorg. Chem. Commun.* **2014**, *50*, 117–121. [CrossRef]

27. Kitos, A.A.; Efthymiou, C.G.; Manos, M.J.; Tasiopoulos, A.J.; Nastopoulos, V.; Escuer, A.; Perlepes, S.P. Interesting copper(II)–assisted transformations of 2-acetylpyridine and 2-benzoylpyridine. *Dalton Trans.* **2016**, *45*, 1063–1077. [CrossRef]

28. Nakamoto, K. *Infrared and Raman Spectra of Inorganic and Coordination Compounds*, 4th ed.; Wiley: New York, NY, USA, 1986; pp. 121–125, 130–139, 254–257.

29. Mylonas-Margaritis, I.; Maniaki, D.; Mayans, J.; Ciammaruchi, L.; Bekiari, V.; Raptopoulou, C.P.; Psycharis, V.; Christodoulou, S.; Escuer, A.; Perlepes, S.P. Mononuclear Lanthanide(III)- Salicylideneaniline Complexes: Synthetic, Structural, Spectroscopic, and Magnetic Studies. *Magnetochemistry* **2018**, *4*, 45. [CrossRef]

30. Mylonas-Margaritis, I.; Mayans, J.; Sakellakou, S.-M.; Raptopoulou, C.P.; Psycharis, V.; Escuer, A.; Perlepes, S.P. Using the Singly Deprotonated Triethanolamine to Prepare Dinuclear Lanthanide(III) Complexes: Synthesis, Structural Characterization and Magnetic Studies. *Magnetochemistry* **2017**, *3*, 5. [CrossRef]

31. Tangoulis, V.; Raptopoulou, C.P.; Terzis, A.; Paschalidou, S.; Perlepes, S.P.; Bakalbassis, E.G. Octanuclearity in Copper(II) Chemistry: Preparation, Characterization, and Magnetochemistry of $[Cu_8(dpk \cdot OH)_8(O_2CCH_3)_4]$ $(ClO_4)_4 \cdot 9H_2O$ (dpk·H_2O = The hydrated, gem-Diol Form of Di-2-pyridyl Ketone). *Inorg. Chem.* **1997**, *36*, 3996–4006. [CrossRef]

32. Lever, A.B.P. *Inorganic Electronic Spectroscopy*, 2nd ed.; Elsevier: Amsterdam, The Netherlands, 1984; pp. 355, 356, 473–478, 507–520.

33. Cotton, F.A.; Wilkinson, G.; Murillo, C.A.; Bochmann, M. *Advanced Inorganic Chemistry*, 6th ed.; Wiley: New York, NY, USA, 1999; pp. 824, 825, 838, 839.

34. Alexopoulou, K.I.; Zagoraiou, E.; Zafiropoulos, T.F.; Raptopoulou, C.P.; Psycharis, V.; Terzis, A.; Perlepes, S.P. Mononuclear anionic octahedral cobalt(III) complexes based on N-salicylidene-o-aminophenol and its derivatives: Synthetic, structural and spectroscopic studies. *Spectrochim. Acta* **2015**, *A136*, 122–130. [CrossRef]

35. Coxall, R.A.; Harris, S.G.; Henderson, D.K.; Parsons, S.; Tasker, P.A.; Winpenny, R.E.P. Inter-ligand reactions: In situ formation of new polydentate ligands. *J. Chem. Soc. Dalton Trans.* **2000**, 2349–2356. [CrossRef]

36. Addison, A.W.; Rao, T.N.; Reedijk, J.; Rijn, J.V.; Verschoor, G.C. Synthesis, structure, and spectroscopic properties of copper(II) compounds containing nitrogen-sulphur donor ligands: The crystal and molecular structure of aqua[1,7-bis(N-methylbenzimidazol-2′-yl)-2,6-dithiaheptane]copper(II) perchlorate. *J. Chem. Soc. Dalton Trans.* **1984**, 1349–1356. [CrossRef]

37. Papatriantafyllopoulou, C.; Stamatatos, T.C.; Wernsdorfer, W.; Teat, S.J.; Tasiopoulos, A.J.; Escuer, A.; Perlepes, S.P. Combining Azide, Carboxylate, and 2-Pyridyloximate Ligands in Transition-Metal Chemistry: Ferromagnetic Ni^{II}_5 Clusters with a Bowtie Skeleton. *Inorg. Chem.* **2010**, *49*, 10486–10496. [CrossRef]

38. Mylonas-Margaritis, I.; Mayans, J.; Perlepe, P.S.; Raptopoulou, C.P.; Psycharis, V.; Alexopoulou, K.I.; Escuer, A.; Perlepes, S.P. Nickel(II) Coordination Clusters Based on N-salicylidene-4-chloro-o-aminophenol: Synthetic and Structural Studies. *Curr. Inorg. Chem.* **2017**, *7*. [CrossRef]

39. Polyzou, C.D.; Lada, Z.G.; Terzis, A.; Raptopoulou, C.P.; Psycharis, V.; Perlepes, S.P. The fac diastereoisomer of tris(2-pyridinealdoximato)cobalt(III) and a cationic cobalt(III) complex containing both the neutral and anionic forms of the ligand: Synthetic, structural and spectroscopic studies. *Polyhedron* **2014**, *79*, 29–36. [CrossRef]

40. Polyzou, C.D.; Koumousi, E.S.; Lada, Z.G.; Raptopoulou, C.P.; Psycharis, V.; Rouzières, M.; Tsipis, A.C.; Mathonière, C.; Clérac, R.; Perlepes, S.P. "Switching on" the single- molecule magnet properties within a series of dinuclear cobalt(III)–dysprosium(III) 2-pyridyloximate complexes. *Dalton Trans.* **2017**, *46*, 14812–14825. [CrossRef]

41. Howson, S.E.; Scott, P. Approaches to the synthesis of optically pure helicates. *Dalton Trans.* **2011**, *40*, 10268–10277. [CrossRef]

42. Piguet, C.; Bernardinelli, G.; Bocquet, B.; Quattropani, A.; Williams, A.F. Self-Assembly of Double and Triple Helices Controlled by Metal Ion Stereochemical Preferences. *J. Am. Chem. Soc.* **1992**, *114*, 7440–7451. [CrossRef]

43. Tuna, F.; Lees, M.R.; Clarkson, G.T.; Hannon, M.J. Readily Prepared Metallo-Supramolecular Triple Helicates Designed to Exhibit Spin-Crossover Behaviour. *Chem. Eur. J.* **2004**, *10*, 5737–5750. [CrossRef]

44. Pelleteret, D.; Clèrac, R.; Mathonière, C.; Harté, E.; Schmitt, W.; Kruger, P.E. Asymmetric spin crossover behaviour and evidence of light-induced excited spin state trapping in a dinuclear iron(II) helicate. *Chem. Commun.* **2009**, 221–223. [CrossRef]

45. Lehn, J.-M. *Supramolecular Chemistry*; VCH: Weinheim, Germany, 1995; p. 150.

46. Steed, J.W.; Atwood, J.L. *Supramolecular Chemistry*; Wiley: New York, NY, USA, 2000; pp. 540–557.

47. *MAGMUN4.1/OW01.exe* is Available as A Combined Package Free of Charge from the Authors. MAGMUN has been Developed by Dr. Zhiqiang Xu, in Collaboration with Prof. L.K. Thompson, and Dr. Oliver Waldmann. Available online: https://www.ucs.mun.ca./~{}lthomp/magmun (accessed on 18 June 2019).

48. Xu, Z.; Thompson, L.K.; Miller, D.O. Dicopper(II) Complexes Bridged by Single N-N Bonds. Magnetic Exchange Dependence on the Rotation Angle between the Magnetic Planes. *Inorg. Chem.* **1997**, *36*, 3985–3995. [CrossRef]

49. Thompson, L.K.; Xu, Z.; Goeta, A.E.; Howard, J.A.K.; Clase, H.J.; Miller, D.O. Structural and Magnetic Properties of Dicopper(II) Complexes of Polydentate Diazine Ligands. *Inorg. Chem.* **1998**, *37*, 3217–3229. [CrossRef]

50. Klaaijsen, F.W.; Blöte, H.W.J.; Dokoupil, Z. Large Single-Ion Anisotropy and Ferromagnetic Intrachain Interaction in four Ni-Compounds of Formula NiX_2L_2 with X= Cl, Br; L= $N_2C_3H_4$, NC_5H_5. *Solid State Commun.* **1974**, *14*, 607–612. [CrossRef]

51. Warner, T.D.; Karnezos, M.; Friedberg, S.A.; Jafarey, S.; Wang, Y.-L. High-temperature series analysis of spin-one uniaxial ferromagnets: $NiMF_6 \cdot 6H_2O$ compounds. *Phys. Rev. B* **1981**, *24*, 2817–2824. [CrossRef]

52. Iio, K.; Nagata, K. Observation of Tetragonal Optical Spectra of Divalent Nickel-Ion in Ba_2NiF_6. *J. Phys. Soc. Jpn.* **1976**, *41*, 1550–1555. [CrossRef]

53. Nguyen, T.N.; Giaquinta, D.M.; zur Loye, H.-C. Synthesis of the New One-Dimentional Compound Sr_3NiPtO_6: Structure and Magnetic Properties. *Chem. Mater.* **1994**, *6*, 1642–1646. [CrossRef]

54. Estes, W.E.; Weller, R.R.; Hatfield, W.E. Preparation and Magnetic Characterization of the Linear-Chain Series $M^{II}L_2Cl_2$ with M= Mn^{2+}, Cu^{2+}, and Ni^{2+} and L= 4-Phenylpyridine. *Inorg. Chem.* **1980**, *19*, 26–31. [CrossRef]

55. Claridge, J.B.; Layland, R.C.; Henley, W.H.; zur Loye, H.-C. Crystal Growth and Magnetic Measurements on Aligned Single Crystals of the Oxides Sr_3NiPtO_6 and Sr_3CuPtO_6. *Chem. Mater.* **1999**, *11*, 1376–1380. [CrossRef]

56. Stamatatos, T.C.; Escuer, A.; Abboud, K.A.; Raptopoulou, C.P.; Perlepes, S.P.; Christou, G. Unusual Structural Types in Nickel Cluster Chemistry from the Use of Pyridyl Oximes: Ni_5, $Ni_{12}Na_2$, and Ni_{14} Clusters. *Inorg. Chem.* **2008**, *47*, 11825–11838. [CrossRef]

57. Alexopoulou, K.I.; Terzis, A.; Raptopoulou, C.P.; Psycharis, V.; Escuer, A.; Perlepes, S.P. Ni^{II}_{20} "Bowls" from the use of Tridentate Schiff Bases. *Inorg. Chem.* **2015**, *54*, 5615–5617. [CrossRef]

58. Carlin, R.L. *Magnetochemistry*; Springer: Berlin, Germany, 1986.

59. *CrystalClear*; ver. 1.40; Rigaku/MSC: The Woodlands, TX, USA, 2005.

60. Sheldrick, G.M. A short history of SHELX. *Acta Crystallogr. A* **2008**, *64*, 112–122. [CrossRef]

61.   Sheldrick, G.M. Crystal structure refinement with SHELXL. *Acta Crystallogr. C* **2015**, *71*, 3–8.
62.   *DIAMOND, Crystal and Molecular Structure Visualization*; ver. 3.1; Crystal Impact: Bonn, Germany, 2018.

 *magnetochemistry*

Article

Relationship between the Coordination Geometry and Spin Dynamics of Dysprosium(III) Heteroleptic Triple-Decker Complexes

Tetsu Sato [1], Satoshi Matsuzawa [2], Keiichi Katoh [1,*], Brian K. Breedlove [1] and Masahiro Yamashita [1,3,4,*]

[1]   Department of Chemistry, Graduate School of Science, Tohoku University, Aramaki-Aza-Aoba, Aoba-ku, Sendai 980-8578, Japan; tetsu.sato.r1@dc.tohoku.ac.jp (T.S.); breedlove@m.tohoku.ac.jp (B.K.B.)

[2]   Institute for Materials Research, Tohoku University, 2–1-1 Katahira, Aoba-ku, Sendai 980-8577, Japan; matsuzawa@imr.tohoku.ac.jp

[3]   Advanced Institute for Materials Research, Tohoku University, 2–1-1 Katahira, Aoba-ku, Sendai 980-8577, Japan

[4]   School of Materials Science and Engineering, Nankai University, Tianjin 300350, China

*   Correspondence: keiichi.katoh.b3@tohoku.ac.jp (K.K.); yamasita@agnus.chem.tohoku.ac.jp (M.Y.); Tel.: +81-22-795-6547 (K.K.); +81-22-795-6544 (M.Y.)

Received: 10 August 2019; Accepted: 22 November 2019; Published: 26 November 2019

Abstract: When using single molecule magnets (SMMs) in spintronics devices, controlling the quantum tunneling of the magnetization (QTM) and spin-lattice interactions is important. To improve the functionality of SMMs, researchers have explored the effects of changing the coordination geometry of SMMs and the magnetic interactions between them. Here, we report on the effects of the octa-coordination geometry on the magnetic relaxation processes of dinuclear dysprosium(III) complexes in the low-temperature region. Mixed ligand dinuclear Dy^{3+} triple-decker complexes [(TPP)Dy(Pc)Dy(TPP)] (**1**), which have crystallographically equivalent Dy^{3+} ions, and [(Pc)Dy(Pc)Dy(TPP)] (**2**), which have non-equivalent Dy^{3+} ions, (Pc^{2-} = phthalocyaninato; TPP^{2-} = tetraphenylporphyrinato), undergo dual magnetic relaxation processes. This is due to the differences in the ground states due to the twist angle (φ) between the ligands. The relationship between the off-diagonal terms and the dual magnetic relaxation processes that appears due to a deviation from D_{4h} symmetry is discussed.

Keywords: Dy^{3+} ion; triple-decker; spin dynamics; octa-coordination geometry

1. Introduction

Rational design and synthesis of single molecular magnets (SMMs) and molecular nanomagnets (MNMs) suitable for quantum information processing (QIP) in quantum computers (QCs) remains difficult [1–7]. Over the past two decades, a wide range of SMMs with controlled spin relaxation processes and high performance have been reported. The charge density distribution of oblate-type lanthanoid ions, (like terbium(III) and dysprosium(III)), strongly improves axial coordination properties of square antiprism (D_{4d}), pentagonal bipyramidal (D_{5h}), and "vent metallocene" type complexes (SMM characteristics have been confirmed for C_1 symmetry) [8–10]. In the D_{4d} ligand field system, the SMM characteristics can be controlled by manipulating the ground state via rotation of the ligands by protonation/deprotonation [11,12], coupling of Tb^{3+}-bisphthalocyaninato (Pc^{2-}) complex with cadmium ions, etc. [13]. Recently, several groups have shown the relationship between octa-coordination environments and the magnetic relaxation processes of Tb^{3+}-Pc^{2-} multiple-decker SMMs along the C_4 rotation axis. From these reports, correlations between the twist angle (φ), ligand field (LF) parameters, and the lowest ground state of Pc^{2-} have been clarified [14]. However, the influence of the coordination environment on the spin relaxation phenomena is not completely understood. In the case of Dy^{3+} ions, the magnetic relaxation mechanism strongly depends on the electronic structure and can sometimes be complicated. In 2003, Ishikawa and coworkers determined that the lowest ground states (J_z) of the $DyPc_2$ complex to be $|J_z| = 13/2$ by using magnetic measurements when φ is 45° and $|J_z| = 11/2$ when φ is 32° [15,16]. In our previous work, we have shown that for some complexes with two crystallographically equivalent Dy^{3+} ions, the ground state is split by the Zeeman effect only if the angle is 45°, resulting in dual relaxation processes. There have been other attempts to elucidate the identities of the dual processes by using theoretical and experimental approaches. Thus, it is important to carefully design the coordination environment around the Dy^{3+} ion in order to investigate the correlation between the magnetic relaxation process and the ground state in detail. In other words, due to the sensitivity toward the coordination environment, Dy^{3+} complexes are easier to control than Tb^{3+} complexes. In this paper, in addition to the dinuclear Dy^{3+} triple-decker complexes we have reported so far, we discuss the relationship between coordination environment and magnetic properties for (TPP)Dy(Pc)Dy(TPP)] (**1**) and [(Pc)Dy(Pc)Dy(TPP)] (**2**) (Pc^{2-} = phthalocyaninato and TPP^{2-} = tetraphenylporphyrinato) with C_4 symmetry.

2. Results and Discussion

2.1. Synthesis and Crystal Strucures

We synthesized the compounds following reported procedures [17]. Single crystal X-ray diffraction analyses on triple-decker complexes **1** and **2** were performed to investigate the coordination environment around the Dy^{3+} ions (Table S1). For **1**, which crystallized in the tetragonal space group $I4/m$, the twist angle (φ) between the outer TPP^{2-} ligand and the inner Pc ligand was determined to be 4°. Therefore, the coordination environment around the Dy^{3+} ions has a slightly distorted square-prism (SP) structure. Since the inner Pc^{2-} ligand acts as a mirror plane in the molecule, the two Dy^{3+} ions are equivalent, and the distance between them was determined to be 3.71 Å. This is a relatively long value among the dinuclear Dy^{3+} triple-decker complexes reported so far. Compound **2** crystallized in the monoclinic space group $C2_1/c$, and the Dy^{3+} ions are inequivalent. φ between the outer and the inner Pc^{2-} ligands was determined to be 39°, meaning that it has a square-antiprism (SAP) structure. φ between the inner Pc^{2-} ligand and the TPP^{2-} ligand was 14°, which is indicative of a highly distorted SAP structure.

Besides the φ between the ligands in the multi-decker type SMM [18], the distance between the coordination isoindole nitrogen atom (N_{iso}) and the metal ions (d), and the angles between the C_4 axis and the direction of Dy^{3+}–N_{iso} (α) affect the properties of these type of SMMs [19]. From a comparison of **1** and **2** with previously studied Dy^{3+} triple-decker complexes **1–6** (Table 1) [20,21], site A of **1** and **2** involving a TPP^{2-} ligand has a small φ. Moreover, distances d_2 and d_3 between the N_{iso} of the Pc^{2-} ligand and the metal ions, as indicated in Figure 1, are greater than those in other complexes. **6** has a large d_2 at the small φ site, and φ decreases with increases in d_2 and d_3. In other words, the TPP^{2-} ligands strongly push the Dy^{3+} ions to the outside of the complex.

The ground states of the Dy^{3+} ions can be estimated from the structure, and theoretical calculations of electrostatic potentials distributed over α for the 4f-shells indicate that the ground state $|J_z|$ is highly dependent on α and d [22]. The ground states of **1** and **2** were estimated from the data in Table 1, and $|J_z| = 11/2$ and 13/2 are the lowest ground states, respectively. However, the $|J_z|$ levels of these dinuclear Dy^{3+} triple-decker complexes are known to mix [19], and intermediate $|J_z|$ levels have the lowest basal order since α is near the magic angle of 54.7° [20,21]. Therefore, φ determines the ground states of the Dy^{3+} ions.

Crystal packing diagrams for **1** and **2** are shown in Figure 2. The intermolecular Dy^{3+}-Dy^{3+} distances along the c axis in **1** and **2** were determined to be 13.61 Å and 11.63 Å, respectively. The molecules of **1** and **2** are rather well-separated from neighboring molecules due to the tetraphenyl groups of the TPP^{2-} ligands and chloroform molecules as crystal solvents (Table S2). Furthermore, PXRD patterns for **1** and **2** at 293 K are slightly different from those simulated from X-ray single crystallographic data for **1** at 120 K because of partial elimination of crystal solvents (Figure S7). From elemental analysis, some of the crystal solvent desorbed. However, the magnetic properties of the same sample remained unchanged after several months, meaning that solvent desorption has little effect on the magnetic properties. The crystal system is the same regardless of whether a solvent is present or absent, meaning that it does not affect the magnetic measurements.

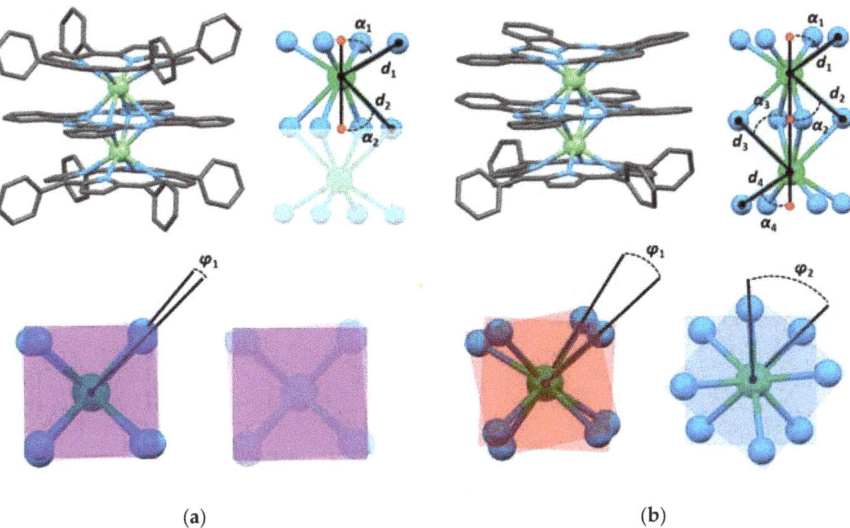

(a) **(b)**

Figure 1. Top and side views of (**a**) [(TPP)Dy(Pc)Dy(TPP)] (**1**) and (**b**) [(Pc)Dy(Pc)Dy(TPP)] (**2**). The bottom of (**a**) and (**b**) are enlargements of the coordination spheres around the Dy^{3+} ions. H atoms and crystal solvents are omitted for clarity. Dy^{3+}: light green, C: grey, and N: light blue.

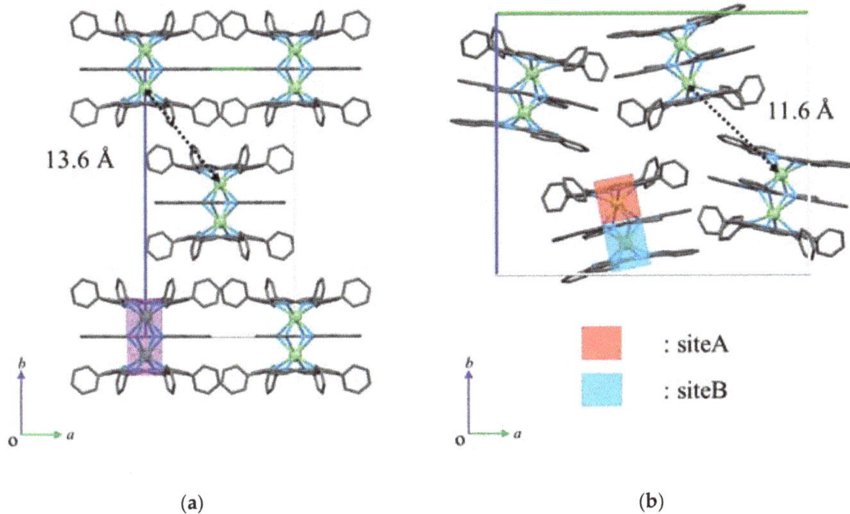

(a) **(b)**

Figure 2. Packing diagrams for (**a**) [(TPP)Dy(Pc)Dy(TPP)] (**1**) and (**b**) [(Pc)Dy(Pc)Dy(TPP)] (**2**). Intermolecular Dy^{3+} distances indicated by black dotted arrows. H atoms and crystal solvents are omitted for clarity. Dy^{3+}: light green, C: grey, and N: light blue.

Table 1. Selected crystallographic data for Dy^{3+}-Pc triple-decker complexes.

Complex	α_1 [°]	α_2 [°]	α_3 [°]	α_4 [°]	d_1 [Å]	d_2 [Å]	d_3 [Å]	d_4 [Å]	φ_A [°]	φ_B [°]
[(TPP)Dy(Pc)Dy(TPP)] **1**	59	46	46	59	2.39	2.67	2.67	2.39	4	4
[(Pc)Dy(Pc)Dy(TPP)] **2**	57	49	45	60	2.35	2.57	2.72	2.37	39	14
[(Pc)Dy(ooPc)Dy(Pc)] **3** [a]	57	48	48	57	2.35	2.59	2.59	2.35	45	45
[(ohPc)Dy(ohPc)Dy(ohPc)] **4** [b]	57	47	47	57	2.35	2.60	2.60	2.35	27	27
[(obPc)Dy(obPc)Dy(obPc)] **5** [c]	57	48	48	57	2.35	2.60	2.60	2.35	32	32
[Dy$_4$] **6** [d]	57	46	49	57	2.33	2.65	2.53	2.37	23	45

[a] ooPc = 2,3,9,10,16,17,23,24-octakis(octyloxy)phthalocyaninato; [b] ohPc^{2-} = 2,3,9,10,16,17,23,24-octakis(hexyloxy)phthalocyaninato; [c] obPc^{2-} = 2,3,9,10,16,17,23,24-octakis(butoxy)phthalocyaninato; [d] [Dy(obPc)$_2$]Dy(Fused-Pc)Dy[Dy(obPc)$_2$] (Fused-Pc^{4-} = bis{72,82,122,132,172,182-hexabutoxytribenzo[g, l, q]-5, 10, 15, 20-tetraazaporphirino}[b, e]benzenato).

2.2. Static Magnetic Properties

The static magnetic susceptibility of **1** and **2** were measured in the *T* range of 1.8–300 K using a superconducting quantum interference device (SQUID) magnetometer. The $\chi_M T$ value at 300 K is consistent with the value expected for the two Dy^{3+} ions ($^6H_{15/2}$, S = 5/2, L = 5, g = 4/3). Curie–Weiss plots for **1** and **2** are shown in Figure 3. Linear approximation was performed over the entire *T* range to obtain values of the Curie (C) (28.50 cm^3 K mol^{-1} (**1**) and 28.20 cm^3 K mol^{-1} (**2**)) and Weiss constants (θ) (−2.33 K (**1**) and −1.97 K (**2**)) (Figures S8 and S9). In $\chi_M T$ versus *T* plots, the values for **1** decreased with a decrease in *T*, reaching a minimum of 19.9 cm^3 K mol^{-1} at 1.8 K, which indicates that magnetic properties of the Dy^{3+} complexes mainly originate from LF effects, such as thermal depopulation of the Stark sublevels [23–25]. As for **2**, there was a slight increase in the $\chi_M T$ values when *T* < 4 K. Since the intermolecular metal distance is sufficiently long [26], the increase is thought to be due to the magnetic dipole interactions between the Dy^{3+} ions in the molecule. All Dy^{3+}-Pc complexes so far reported exhibit similar ferromagnetic behavior. However, the behavior of the $\chi_M T$ values for **1** is different from the other Dy^{3+} triple-decker complexes, and an increase in the $\chi_M T$ values was not observed in the low *T* region. Although the fact that the intermolecular distance is similar for each complex, the difference in LF parameters of the Dy^{3+} ions has a dramatic effect. In other words, the increase in the off-diagonal terms and the change in the LF splitting energy along with the change in symmetry are important factors affecting the magnetic behavior. Fitting of the data was performed using the PHI program with reported LF parameters [27]. However, we could not obtain consistent results for **1** and **2** because their ground states are complicated due to mixing of the off-diagonal terms.

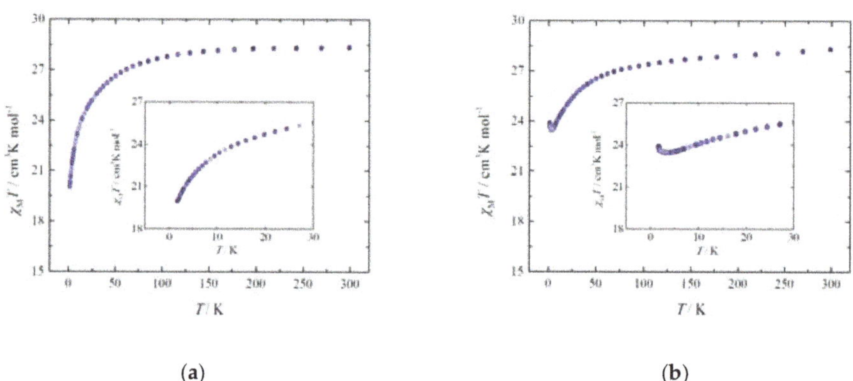

(a)   (b)

Figure 3. *T* dependence of $\chi_M T$ measured on powder samples of (**a**) **1** and (**b**) **2** in the *T* range of 1.8–300 K in an H_{dc} of 0.5 kOe. The inset is a magnified plot of 1.8–30 K.

In order to confirm the magnetic anisotropy of the molecule, *T* dependence of *MH* were performed (Figure 4). For both complexes, the magnetization did not saturate up to 70 kOe. However, splitting of the M versus HT^{-1} plot occurred, indicating that not only depopulation occurred but also both complexes had large magnetic anisotropies. In addition, butterfly-type hysteresis was not observed during the MH measurements at 1.8 K. When uniaxial anisotropy is strong, the saturation magnetization value (M_s) is expressed as $M_s = 1/2 \times g^*(z) \times \widetilde{S}$ where $\widetilde{S} = 1/2$ [28]. So, the M_s values of **1** and **2** were $M_s = 8.6\ \mu_B$ and $M_s = 7.4\ \mu_B$, calculated using $|J_z| = 13/2$ ($g^*(x) = g^*(y) = 0$, $g^*(z) = 17.3$) and $11/2$ ($g^*(x) = g^*(y) = 0$, $g^*(z) = 14.7$), respectively. Although the measured values are larger than the calculated values (**1**: 13.6 μ_B, **2**: 13.7 μ_B), they are smaller than the effective magnetic moments (μ_{eff}) of two Dy^{3+} ions ($\mu_{\text{eff}} = 21\ \mu_B$).

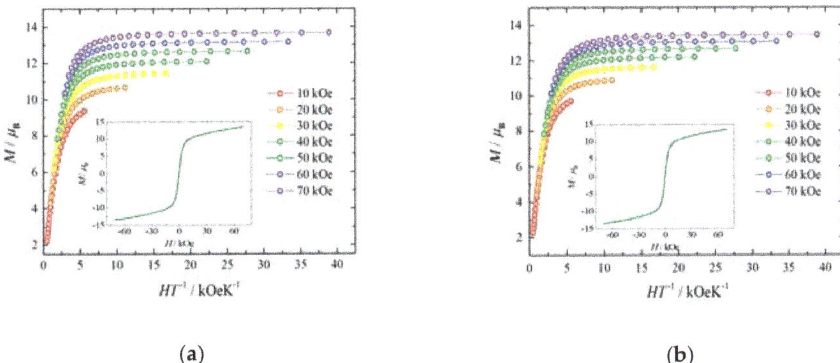

(a)   (b)

Figure 4. Magnetic field (H) dependence of the magnetization (M) for powder samples of (a) **1** and (b) **2** in the T range of 1.8–20 K.

2.3. Dynamic Magnetic Properties

In order to investigate the magnetic relaxation processes, alternating current (ac) magnetic susceptibility measurements were performed on powder samples of **1** and **2**. For **2**, the χ_M'' values were frequency (v) dependent in the range of 0.1–1000 Hz in a zero magnetic field (Figures S10–S13), whereas for **1**, they were not. To clarify these differences, ac magnetic susceptibility measurements were performed in different magnetic fields at 1.8 K. When H was in the range of 0–5 kOe, the magnetization of both complexes underwent dual relaxation processes (Figure 5). As shown in Figure 6, the magnetic relaxation times (τ) calculated from the χ_M'' versus v plot on the low v side increased monotonically with an increase in H, and the τ values calculated from the high v side reached a local maximum in specific applied H_{dc}. These results indicate that the maximum H_{dc} suppresses QTM and promotes a direct process [29], and the results can be reproduced using a mixture of QTM, direct and Raman processes with Equation (1) (Table S3):

$$\tau^{-1} = AH^nT + \frac{B_1}{1 + B_2H^2} + D. \tag{1}$$

In previous work, we have reported that compound **3** exhibited dual magnetic relaxation processes when H_{dc} was larger than 2.5 kOe [14]. To understand the details of the dual magnetic relaxation dynamics of **1** and **2**, we analyzed the ratio of relaxation time ρ (= τ_2/τ_1). The ρ values of 1 correspond to the occurrence of a single relaxation process in the Dy^{3+} SMM system. If the value is large enough (>1000), dual magnetic relaxation processes are observed separately [30]. Since the splitting of relaxation is observed at 0.25 kOe for **1** and **2**, it can be said that it responds more sensitively to H_{dc}. This is another indication that the ground states of Dy^{3+} ions are more complicated.

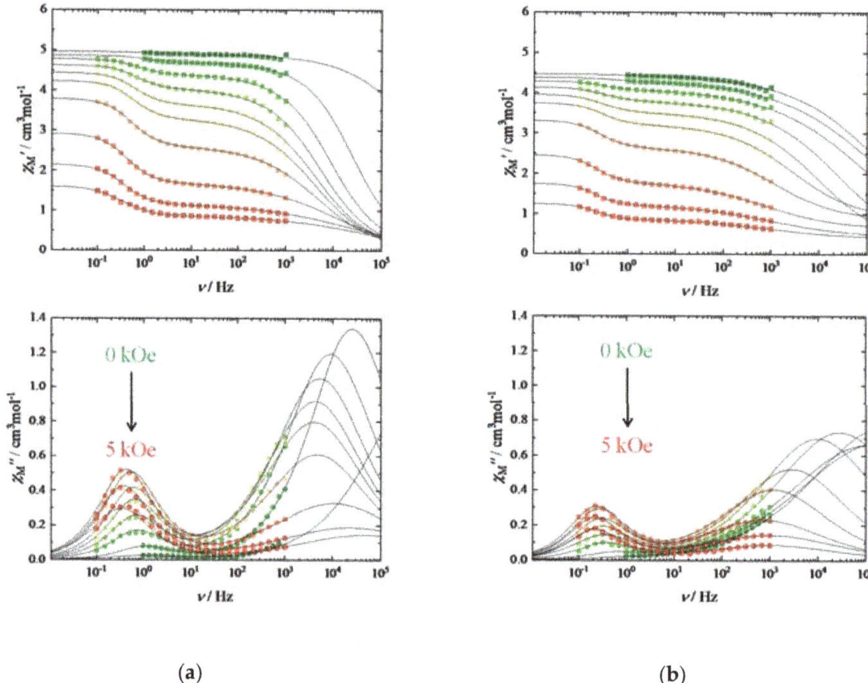

(a)   (b)

Figure 5. ν dependence of the real (χ_M') and imaginary (χ_M'') parts of the ac susceptibilities of (**a**) **1** and (**b**) **2** in H_{dc} of 0–5 kOe at 1.8 K. Black solid lines were fitted by using an extended Debye model to obtain τ. Argand plots are in the supporting information (Figures S16 and S17)

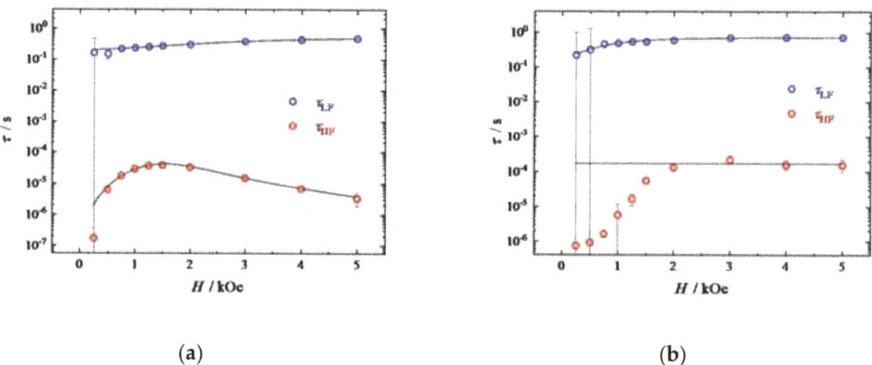

(a)   (b)

Figure 6. (**a**) τ versus H for **1** at 1.8 K obtained from the least-squares fitting using an extended Debye model. τ from the high frequency region have the maxima each other. (**b**) τ versus H for **2**. The data were fitted using a mixture of as the quantum tunneling of the magnetization (QTM), direct and Raman processes with the parameters listed in the SI.

ν dependences of the χ_M'' values of **1** and **2** were measured in the range of 1–1000 Hz in various H_{dc}, and a split in the χ_M'' values was observed below 8 and 20 K, respectively. In a χ_M'' versus T plot, a peak top was observed in the T range below 2 K, indicating that the magnetic moment was not frozen or that a different relaxation processes, like QTM, was dominate. We used the Kramers–Kronig equation [31–35], which infers the pre-exponential factor τ_0 and the activation barrier U_{eff} from the χ_M''/χ_M' versus T (2.5–4 K) plot, to fit the data:

$$\chi_M''/\chi_M' = \omega\tau, \tag{2}$$

$$\chi_M''/\chi_M' = \omega\tau_0 + \exp(U_{\text{eff}}/T), \tag{3}$$

$$\ln(\chi_M''/\chi_M') = \ln(\omega\tau_0) + U_{\text{eff}}/T. \tag{4}$$

From a χ_M''/χ_M' plot for **1**, U_{eff} was determined to be about 8.1 cm^{-1}, and $\tau_0 \approx 10^{-7}$ s. For **2**, U_{eff} was determined to be about 2.7 cm^{-1}, and $\tau_0 \approx 10^{-6}$ s (Figures S14 and S15). The small τ_0 indicates, that the contribution from an Orbach process becomes small. Figure 7 shows the relationship between U_{eff} and φ [15,16]. From this figure, as φ decreases from 45°, the activation energy tends to decrease. This is because a contribution from the off-diagonal LF terms promotes QTM, and during the conversion from SAP to SP geometries. Thus, B_2^0 becomes smaller, and off-diagonal terms B_4^4 and B_6^4 become larger, resulting in a narrower U_{eff}. It is thought that a small φ has a negative effect on the activation barrier. In other words, the small φ of **1** and **2** cause a decrease in the activation barrier.

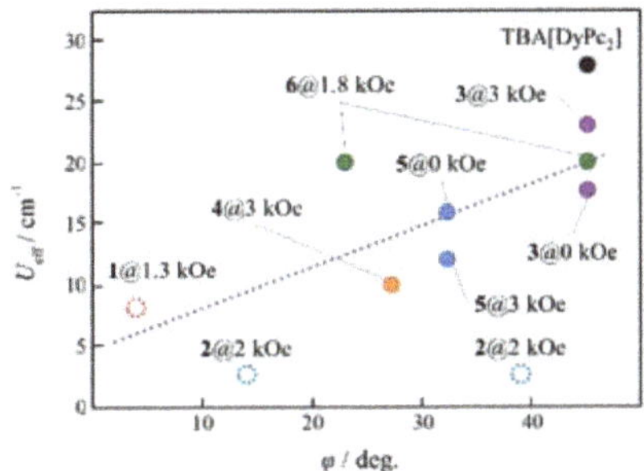

Figure 7. U_{eff} versus φ plots for related Dy^{3+}-Pc single molecule magnets (SMMs). Blue dotted lines are guides only. Dotted circles indicate the values for **1** and **2**, and the color dots indicate **3–6** in H_{dc}. Since **2** and **6** have two different Dy^{3+} sites, the distribution of which could not be separated distribution, both φ values are displayed.

To investigate the magnetic relaxation properties of **1** and **2** at low T, ν dependence measurements were performed in an applied H_{dc} in the T range of 1.8–4.5 K. An Argand plot for both complexes showed that a dual magnetic relaxation process occurred. τ for each component was calculated by fitting the imaginary component of the ac magnetic susceptibility with the extended Debye model (Equations S2 and S3, Figure 8), and using those values, an Arrhenius plot was obtained (Figure 9). From a fitting with Equation (5) on the values for **1**, τ obtained from the low ν side indicates that a QTM process independent of T occurs, and that obtained from the high ν side is proportional to T^{-9}, meaning that it is a Raman process (Table S4). However, the fitting of the data for **1** is not accurate due to large deviation in the τ values. In particular, when τ values at $T > 2.5$ K, the spin dynamics of **1** could not be determined. On the other hand, for **2**, τ obtained from the low ν side is proportional to $T^{-1.7}$, and that obtained from the high ν side indicates that a direct process occurs. For both complexes, the lowest ground state estimated from the crystal structure is expected to contain a large amount of mixing. Therefore, even when a complex has crystallographically equivalent Dy^{3+} ions, dual relaxation processes occur between mixed states in an applied magnetic field. In the case of **1**, the D_{4h} symmetry around the Dy^{3+} ion has a large effect on the off-diagonal term, causing QTM to be dominant. In addition, T is proportional to n $\approx$ −2 which could be reproduced using a PB process to fit the data.

However, since Raman processes can involve acoustic-optical phonons (n = 1–6) [18], it is difficult to separate each contribution due to the complicated ground state of **2**.

$$\tau^{-1} = AH^n T + C T^m + \tau_{QTM}^{-1}. \tag{5}$$

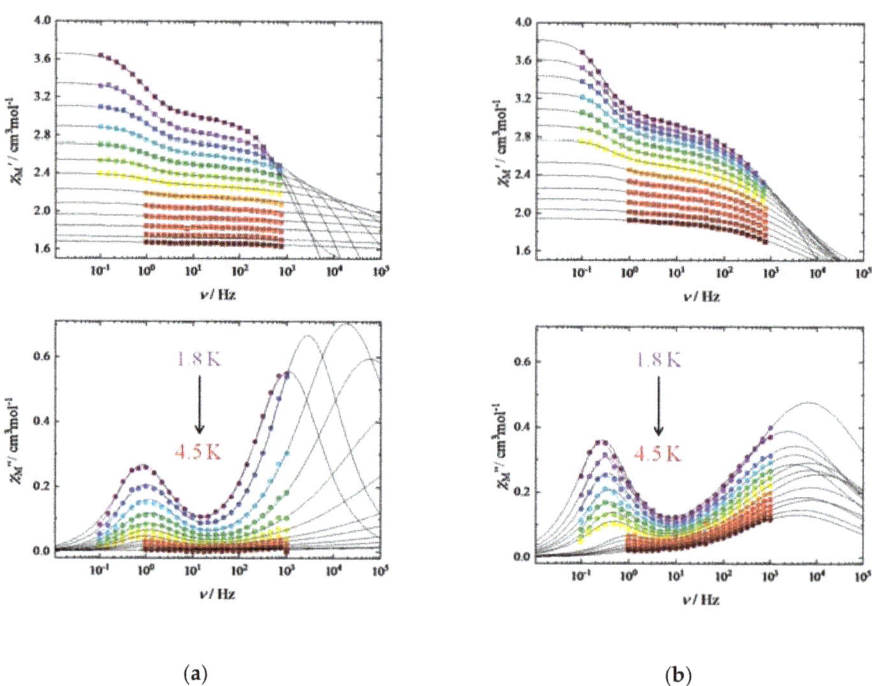

(a)

(b)

Figure 8. ν dependence of the out of phase (χ_M'') parts of the ac magnetic susceptibilities of **1** (a) and **2** (b) measured in the T range of 1.8–4.5 K in H_{dc}. Black solid lines were fitted by using an extended Debye model to obtain τ. Argand plots are located in the supporting information (Figures S18 and S19).

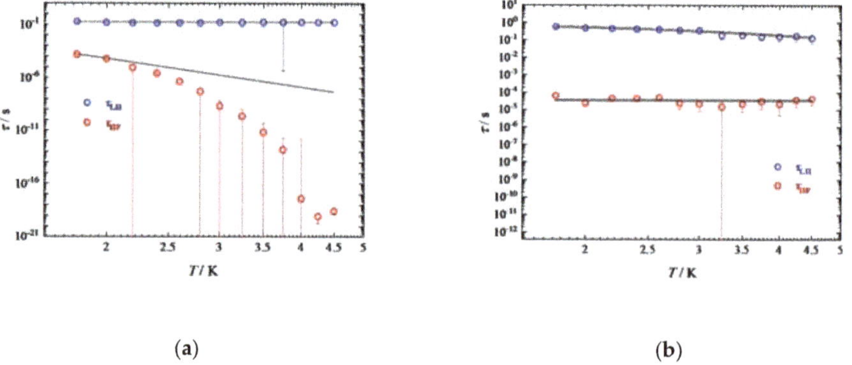

(a)

(b)

Figure 9. An Arrhenius plot for (a) **1** and (b) **2**, for which the τ values were obtained from χ_M'' versus ν plots in H_{dc} of 1.3 and 2 kOe, respectively, in the T range of 1.8–4.5 K. The blue circles indicate the τ values from the low ν region and red circles indicate those from the high ν region. Black solid lines were fitted by using Equation (5).

3. Materials and Methods

3.1. Synthesis

Solvents were used without further purification. Dy(acac)$_3$·4H$_2$O and the free ligand were purchased from TCI Tokyo Chemical Industry Co., LTD, Tokyo, Japan. Dy(acac)$_3$·4H$_2$O (180 mg, 0.40 mmol) and H$_2$TTP (tetraphenylporphyrin) (150 mg, 0.25 mmol) were added to dry 1,2,4-trichlorobenzene (40 mL). The solution was refluxed under nitrogen for 4 h. After cooling, Li$_2$Pc (158 mg, 0.60 mmol) was added to the mixture. Then, the solution was refluxed for 12 h. After cooling, the reaction mixture was added to n-hexane (500 mL). The obtained solid was purified by using column chromatography on silica gel with chloroform as the eluent. [(TTP)Dy(Pc)Dy(TTP)] (**1**) was obtained from a deep brownish red fraction, which was the first fraction, by removing the solvent (16%), and [(Pc)Dy(Pc)Dy(TTP)] (**2**) was obtained from the dark green second fraction (34%). Spectroscopic data used for characterization are described in the SI (Figures S5 and S6). Column chromatography (C-200 silica gel, Wako and Sephadex G-10, Pharmacia Biotech) was used to remove the remaining impurities. Dark red block crystals of **1** were obtained from chloroform/n-hexane (27 mg). ESI-MS: m/z (%): 2062.47242 (100) [M−1$^+$] (Figures S1 and S2); elemental analysis calcd (%) for C$_{120}$H$_{72}$N$_{16}$Dy$_2$·4CHCl$_3$: C 58.62, H 3.02, N 8.82; found: C 60.03, H 3.21, N 8.89. Black fine needle crystals of **2** were obtained from chloroform/n-hexane (83 mg). ESI-MS: m/z (%): 1962.38956 (100) [M$^+$] (Figure S3 and S4); elemental analysis calcd (%) for C$_{108}$H$_{60}$N$_{20}$Dy$_2$·CHCl$_3$: C 60.01, H 2.84, N 12.72; found: C 62.91, H 3.28, N 13.36. The results of the elemental analysis for **2** indicates desorption of some of the CHCl$_3$ molecules compared with the number of CHCl$_3$ molecules calculated from the crystal structure.

3.2. Physical Measurements

Elemental analyses were conducted on a PerkinElmer 240C elemental analyzer (PerkinElmer, Waltham, MA, USA) at the Research and Analytical Centre for Giant Molecules, Tohoku University. UV-Vis-NIR spectra were acquired using CHCl$_3$ solution on a Shimadzu UV-3100pc (Shimadzu, Kyoto, Japan). IR spectroscopy was performed on ATR method on FT/IR-4200 spectrometer at 298 K. Magnetic susceptibility measurements were conducted on a Quantum Design SQUID magnetometer MPMS-XL and MPMS-3 (Quantum Design, San Diego, CA, USA) in the T and dc field ranges of 1.8–300 K and ±50 kOe, respectively. AC measurements were performed in the frequency range of 0.1–1000 Hz with an ac field amplitude of 3 Oe. A polycrystalline sample suspended in n-eicosane was used for the measurements. Crystallographic data for **1** and **2** were collected at 120 K on a Rigaku Saturn724+ CCD Diffractometer (Rigaku, Tokyo, Japan) with graphite-monochromated Mo Kα radiation (λ = 0.71075 Å) produced using a VariMax microfocus X-ray rotating anode source. Single crystals with dimensions of 0.10 × 0.07 × 0.01 mm^3 (**1**) and 0.15 × 0.20 × 0.04 mm^3 (**2**) were used. Data processing was conducted using the Crystal Clear crystallographic software package [36]. The structures were solved by using direct methods using SIR-92 [37]. Refinement was carried out using the Yadokari-XG package [38] and SHELXT. The non-Hydrogen atoms were refined anisotropically using weighted full-matrix least squares on F. Hydrogen atoms attached to the carbon atoms were fixed using idealized geometries and refined using a riding model. CCDC 1940003 for **1** and 1940004 for **2**. Powder X-ray diffraction was conducted on a Bruker AXS D2 phaser (Bruker Corporation Billerica, MA, USA).

4. Conclusions

Complexes **1** and **2** were synthesized similar to previously reported Dy^{3+}-Pc complexes with a TPP^{2-} ligand. The TPP^{2-} ligands induce a smaller twist angle (φ) between the ligands than those in the previous complexes due to the effects of steric repulsion from the phenyl group. Although **1** has D_{4h} symmetry due to the two TPP^{2-} ligands, **2** has lower symmetry due to having only one TPP^{2-}. From dc magnetic measurements on both complexes, the $\chi_M T$ values decreased due to depopulation. Measurement of M-HT^{-1} indicated uniaxial anisotropy, but it is smaller than the expected values for a pseudospin model. No hysteresis opening at 1.8 K was observed, suggesting a mixture of ground states, which is consistent with the estimation of the ground state using the value of α obtained from the crystal structure data. In addition, dual slow magnetization relaxation was observed for both complexes from ac magnetic susceptibility measurements in an applied H_{dc}. U_{eff} calculated by using the Kramers–Kronig equation is very small and corresponded to the tendency of previous triple-decker compounds to decrease with decrease of φ. From above the results, **1** and **2** are field-induced SMMs. For **1**, QTM and Raman processes occur due to the symmetry of D_{4h}, whereas for **2**, mixed relaxation (Raman, PB) and QTM processes occur. The contributions of the Raman and PB processes must be clarified. The magnetic processes involve spin relaxation in mixed ground states, and the off-diagonal term is dominant. This is different from conventional dinuclear Dy^{3+} complexes. Multiple relaxation processes could be turned on by adjusting φ to 4° (**1**) and 14° (**2**), and this can be used to prepare functional SMMs whose characteristics can be switched on and off by changing φ.

Supplementary Materials: The following are available online at http://www.mdpi.com/2312-7481/5/4/65/s1. Figure S1: ESI-MS spectrum of **1** in $CHCl_3$. The peak at 2062.47242 corresponds to [M-1^+]. Figure S2: Experimental (top) and simulated (bottom) ESI-MS spectra of **1** in $CHCl_3$. The peak at 2062.47242 corresponds to [M−1^+]. Figure S3: ESI-MS spectrum of **2** in $CHCl_3$. The peak at 1962.38956 corresponds to [M^+]. Figure S4: Experimental (top) and simulated (bottom) ESI-MS spectra of **2** in $CHCl_3$. The peak at 1962.38956 corresponds to [M^+]. Figure S5: IR spectrum for **1** (top) and **2** (bottom) by using an ATR method at 298 K. Figure S6: UV-vis-NIR spectra for **1** (top) and **2** (bottom) in $CHCl_3$ (5.1×10^{-3} (**1**), and 4.7×10^{-3} (**2**)) at 298 K. Figure S7: PXRD patterns for 1 (top) and 2 (bottom). Figure S8: Curie–Weiss plot for **1**. Linear approximation is performed over the entire T range, from which the values of Curie constant (C) (28.50 cm^3 K mol^{-1}) and Weiss constant (θ) (−2.33 K) were obtained. Figure S9: Curie–Weiss plot for **2**. Linear approximation is performed over the entire T range, from which the values of Curie constant (C) (28.20 cm^3 K mol^{-1}) and Weiss constant (θ) (−1.97 K) were obtained. Figure S10: Frequency (ν) and temperature (T) dependences of the (**a**) in-phase (χ_M') and (**b**) out-of-phase (χ_M'') ac magnetic susceptibilities of **1** in 0 kOe. Figure S11: Frequency (ν) and temperature (T) dependences of the (**a**) in-phase (χ_M') and (**b**) out-of-phase (χ_M'') ac magnetic susceptibilities of **2** in 0 kOe. Figure S12: Frequency (ν) and temperature (T) dependences of the (**a**) in-phase (χ_M') and (**b**) out-of-phase (χ_M'') ac magnetic susceptibilities of **1** in 1.3 kOe. Figure S13: Frequency (ν) and temperature (T) dependences of the (**a**) in-phase (χ_M') and (**b**) out-of-phase (χ_M'') ac magnetic susceptibilities of **2** in 2 kOe. Figure S14: χ_M''/χ_M' versus T (2.5–4 K) plot for **1**. Figure S15: χ_M''/χ_M' versus T (2.5–4 K) plot for **2**. Figure S16: Argand plots (χ_M'' versus χ_M') for **1** at 1.8 K in several dc magnetic fields (0-5 kOe). Black solid lines were guides for eye. Figure S17: Argand plots (χ_M'' versus χ_M') for **2** at 1.8 K in several dc magnetic fields in the range of 0–5 kOe. Black solid lines were guides for eye. Figure S18: Argand plots (χ_M'' versus χ_M') for **1** in 1.3 kOe in the T range of 1.8–4.5 K. Black solid lines are guides for the eye. Figure S19: Argand plots (χ_M'' versus χ_M') for **1** in 2 kOe field in the T range of 1.8–4.5 K. Black solid lines are guides for the eye. Table S1: Selected crystallographic data for **1** and **2**. Table S2: Selected crystallographic data for **1** and **2**. Table S3: Parameters for fitting the τ verses H plots. Table S4: Parameters of fitting for τ verses T plot.

Author Contributions: Conceptualization, K.K.; Data curation, K.K.; Formal analysis, T.S. and S.M.; Funding acquisition, K.K. and M.Y.; Investigation, T.S. and S.M.; Methodology, K.K.; Project administration, K.K. and M.Y.; Supervision, K.K.; Validation, K.K.; Visualization, B.K.B.; Writing—original draft, T.S.; Writing—review & editing, K.K., B.K.B and M.Y.

Funding: This work was supported by a Grant-in-Aid for Scientific Research (S) (grant no. 20225003), Grant-in-Aid for Young Scientists (B) (grant no. 24750119), Grant-in-Aid for Scientific Research (C) (grant no. 15K05467) from the Ministry of Education, Culture, Sports, Science, Technology, Japan (MEXT), CREST (JPMJCR12L3) from JST, a Grant-in-aid for JSPS fellows from the Japan Society for the Promotion of Science (JSPS) (25·2441), and Tohoku University Division for International Advanced Research and Education (DIARE). M.Y. is grateful for the support from the 111 project (B18030) from China. S.M. acknowledges the support by MD-program of Tohoku Univ.

Conflicts of Interest: The authors declare no conflict of interest.

References

1. Sessoli, R.; Gatteschi, D.; Caneschi, A. Magnetic bistability in a metal-ion cluster. *Nature* **1993**, *363*, 141–143. [CrossRef]

2. Wernsdorfer, W.; Sessoli, R. Quantum Phase Interference and Parity Effects in Magnetic Molecular Clusters. *Science* **1999**, *284*, 133–136. [CrossRef] [PubMed]

3. Choi, K.Y.; Wang, Z.; Nojiri, H.; Van Tol, J.; Kumar, P.; Lemmens, P.; Bassil, B.S.; Kortz, U.; Dalal, N.S. Coherent manipulation of electron spins in the (Cu3) spin triangle complex impregnated in nanoporous silicon. *Phys. Rev. Lett.* **2012**, *108*, 1–5. [CrossRef] [PubMed]

4. Vincent, R.; Klyatskaya, S.; Ruben, M.; Wernsdorfer, W.; Balestro, F. Electronic read-out of a single nuclear spin using a molecular spin transistor. *Nature* **2012**, *488*, 357–360. [CrossRef]

5. Collauto, A.; Mannini, M.; Sorace, L.; Barbon, A.; Brustolon, M.; Gatteschi, D. A slow relaxing species for molecular spin devices: EPR characterization of static and dynamic magnetic properties of a nitronyl nitroxide radical. *J. Mater. Chem.* **2012**, *22*, 22272–22281. [CrossRef]

6. Wedge, C.J.; Timco, G.A.; Spielberg, E.T.; George, R.E.; Tuna, F.; Rigby, S.; McInnes, E.J.L.; Winpenny, R.E.P.; Blundell, S.J.; Ardavan, A. Chemical engineering of molecular qubits. *Phys. Rev. Lett.* **2012**, *108*, 1–5. [CrossRef]

7. Aguilà, D.; Barrios, L.A.; Velasco, V.; Roubeau, O.; Repollés, A.; Alonso, P.J.; Sesé, J.; Teat, S.J.; Luis, F.; Aromí, G. Heterodimetallic [LnLn'] lanthanide complexes: Toward a chemical design of two-qubit molecular spin quantum gates. *J. Am. Chem. Soc.* **2014**, *136*, 14215–14222. [CrossRef]

8. Long, J.; Selikhov, A.N.; Mamontova, E.; Lyssenko, K.A.; Guari, Y.; Larionova, J.; Trifonov, A.A. Single-molecule magnet behaviour in a Dy(iii) pentagonal bipyramidal complex with a quasi-linear Cl-Dy-Cl sequence. *Dalt. Trans.* **2019**, *48*, 35–39. [CrossRef]

9. Guo, F.S.; Day, B.M.; Chen, Y.C.; Tong, M.L.; Mansikkamäki, A.; Layfield, R.A. A Dysprosium Metallocene Single-Molecule Magnet Functioning at the Axial Limit. *Angew. Chem. Int. Ed.* **2017**, *56*, 11445–11449. [CrossRef]

10. Day, B.M.; Guo, F.S.; Layfield, R.A. Cyclopentadienyl Ligands in Lanthanide Single-Molecule Magnets: One Ring to Rule Them All? *Acc. Chem. Res.* **2018**, *51*, 1880–1889. [CrossRef]

11. Tanaka, D.; Inose, T.; Tanaka, H.; Lee, S.; Ishikawa, N.; Ogawa, T. Proton-induced switching of the single molecule magnetic properties of a porphyrin based Tb^{III} double-decker complex. *Chem. Commun.* **2012**, *48*, 7796–7798. [CrossRef] [PubMed]

12. Horii, Y.; Horie, Y.; Katoh, K.; Breedlove, B.K.; Yamashita, M. Changing Single-Molecule Magnet Properties of a Windmill-Like Distorted Terbium(III) α-Butoxy-Substituted Phthalocyaninato Double-Decker Complex by Protonation/Deprotonation. *Inorg. Chem.* **2018**, *57*, 565–574. [CrossRef] [PubMed]

13. Horii, Y.; Katoh, K.; Yasuda, N.; Breedlove, B.K.; Yamashita, M. Effects of f-f interactions on the single-molecule magnet properties of terbium(iii)-phthalocyaninato quintuple-decker complexes. *Inorg. Chem.* **2015**, *54*, 3297–3305. [CrossRef] [PubMed]

14. Katoh, K.; Aizawa, Y.; Morita, T.; Breedlove, B.K.; Yamashita, M. Elucidation of Dual Magnetic Relaxation Processes in Dinuclear Dysprosium(III) Phthalocyaninato Triple-Decker Single-Molecule Magnets Depending on the Octacoordination Geometry. *Chem. A Eur. J.* **2017**, *23*, 15377–15386. [CrossRef] [PubMed]

15. Ishikawa, N.; Sugita, M.; Okubo, T.; Tanaka, N.; Iino, T.; Kaizu, Y. Determination of ligand-field parameters and f-electronic structures of double-decker bis(phthalocyaninato)lanthanide complexes. *Inorg. Chem.* **2003**, *42*, 2440–2446. [CrossRef] [PubMed]

16. Ishikawa, N.; Iino, T.; Kaizu, Y. Interaction between f-electronic systems in dinuclear lanthanide complexes with phthalocyanines. *J. Am. Chem. Soc.* **2002**, *124*, 11440–11447. [CrossRef]

17. Tran-Thi, T.H.; Mattioli, T.A.; Chabach, D.; Cian, A.D.; Weiss, R. Hole localization or delocalization? An optical, raman, and redox study of lanthanide porphyrin-phthalocyanine sandwich-type heterocomplexes. *J. Phys. Chem.* **1994**, *98*, 8279–8288. [CrossRef]

18. Zhang, P.; Guo, Y.N.; Tang, J. Recent advances in dysprosium-based single molecule magnets: Structural overview and synthetic strategies. *Coord. Chem. Rev.* **2013**, *257*, 1728–1763. [CrossRef]

19. Sorace, L.; Benelli, C.; Gatteschi, D. Lanthanides in molecular magnetism: Old tools in a new field. *Chem. Soc. Rev.* **2011**, *40*, 3092–3104. [CrossRef]

20. Katoh, K.; Morita, T.; Yasuda, N.; Wernsdorfer, W.; Kitagawa, Y.; Breedlove, B.K.; Yamashita, M. Tetranuclear Dysprosium(III) Quintuple-Decker Single-Molecule Magnet Prepared Using a π-Extended Phthalocyaninato Ligand with Two Coordination Sites. *Chem. A Eur. J.* **2018**, *24*, 15522–15528. [CrossRef]

21. Katoh, K.; Horii, Y.; Yasuda, N.; Wernsdorfer, W.; Toriumi, K.; Breedlove, B.K.; Yamashita, M. Multiple-decker phthalocyaninato dinuclear lanthanoid(iii) single-molecule magnets with dual-magnetic relaxation processes. *Dalt. Trans.* **2012**, *41*, 13582–13600. [CrossRef] [PubMed]

22. Liu, J.L.; Chen, Y.C.; Tong, M.L. Symmetry strategies for high performance lanthanide-based single-molecule magnets. *Chem. Soc. Rev.* **2018**, *47*, 2431–2453. [CrossRef] [PubMed]

23. Bi, Y.; Guo, Y.N.; Zhao, L.; Guo, Y.; Lin, S.Y.; Jiang, S.D.; Tang, J.; Wang, B.W.; Gao, S. Capping ligand perturbed slow magnetic relaxation in dysprosium single-ion magnets. *Chem. A Eur. J.* **2011**, *17*, 12476–12481. [CrossRef] [PubMed]

24. Damjanovic, M.; Katoh, K.; Yamashita, M.; Enders, M. Combined NMR analysis of huge residual dipolar couplings and pseudocontact shifts in terbium(III)-phthalocyaninato single molecule magnets. *J. Am. Chem. Soc.* **2013**, *135*, 14349–14358. [CrossRef] [PubMed]

25. Zou, L.; Zhao, L.; Chen, P.; Guo, Y.N.; Guo, Y.; Li, Y.H.; Tang, J. Phenoxido and alkoxido-bridged dinuclear dysprosium complexes showing single-molecule magnet behaviour. *Dalt. Trans.* **2012**, *41*, 2966–2971. [CrossRef]

26. Katoh, K.; Breedlove, B.K.; Yamashita, M. Symmetry of octa-coordination environment has a substantial influence on dinuclear TbIII triple-decker single-molecule magnets. *Chem. Sci.* **2016**, *7*, 4329–4340. [CrossRef]

27. Chilton, N.F.; Anderson, R.P.; Turner, L.D.; Soncini, A.; Murray, K.S. PHI: A powerful new program for the analysis of anisotropic monomeric and exchange-coupled polynuclear d- and f-block complexes. *J. Comput. Chem.* **2013**, *34*, 1164–1175. [CrossRef]

28. Moreno Pineda, E.; Chilton, N.F.; Marx, R.; Dörfel, M.; Sells, D.O.; Neugebauer, P.; Jiang, S.D.; Collison, D.; Van Slageren, J.; McInnes, E.J.L.; et al. Direct measurement of dysprosium(III)···dysprosium(III) interactions in a single-molecule magnet. *Nat. Commun.* **2014**, *5*, 1–7. [CrossRef]

29. Ding, Y.S.; Yu, K.X.; Reta, D.; Ortu, F.; Winpenny, R.E.P.; Zheng, Y.Z.; Chilton, N.F. Field- and temperature-dependent quantum tunnelling of the magnetisation in a large barrier single-molecule magnet. *Nat. Commun.* **2018**, *9*, 1–10. [CrossRef]

30. Guo, Y.N.; Xu, G.F.; Guo, Y.; Tang, J. Relaxation dynamics of dysprosium(iii) single molecule magnets. *Dalt. Trans.* **2011**, *40*, 9953–9963. [CrossRef]

31. Gass, I.A.; Moubaraki, B.; Langley, S.K.; Batten, S.R.; Murray, K.S. A π-π 3D network of tetranuclear μ2/μ3- carbonato Dy(iii) bis-pyrazolylpyridine clusters showing single molecule magnetism features. *Chem. Commun.* **2012**, *48*, 2089–2091. [CrossRef] [PubMed]

32. Ferrando-Soria, J.; Cangussu, D.; Eslava, M.; Journaux, Y.; Lescouëzec, R.; Julve, M.; Lloret, F.; Pasán, J.; Ruiz-Pérez, C.; Lhotel, E.; et al. Rational enantioselective design of chiral heterobimetallic single-chain magnets: Synthesis, crystal structures and magnetic properties of oxamato-bridged MIICuII chains (M = Mn, Co). *Chem. A Eur. J.* **2011**, *17*, 12482–12494. [CrossRef] [PubMed]

33. Bartolomé, J.; Filoti, G.; Kuncser, V.; Schinteie, G.; Mereacre, V.; Anson, C.E.; Powell, A.K.; Prodius, D.; Turta, C. Magnetostructural correlations in the tetranuclear series of {Fe3 LnO2} butterfly core clusters: Magnetic and Mössbauer spectroscopic study. *Phys. Rev. B Condens. Matter Mater. Phys.* **2009**, *80*, 1–16. [CrossRef]

34. Luis, F.; Bartolomé, J.; Fernández, J.; Tejada, J.; Hernández, J.; Zhang, X. Thermally activated and field-tuned tunneling inAc studied by ac magnetic susceptibility. *Phys. Rev. B Condens. Matter Mater. Phys.* **1997**, *55*, 11448–11456. [CrossRef]

35. Nakanishi, R.; Yatoo, M.A.; Katoh, K.; Breedlove, B.K.; Yamashita, M. Dysprosium Acetylacetonato Single-molecule magnet encapsulated in Carbon Nanotubes. *Materials (Basel)* **2017**, *10*, 7. [CrossRef] [PubMed]

36. Rigaku Corporation. *Crystal Clear-SM, 1.4.0 SP1*; Rigaku Corporation: Tokyo, Japan, 17 April 2008.

37. Altomare, A.; Burla, M.C.; Camalli, M.; Cascarano, G.L.; Giacovazzo, C.; Guagliardi, A.; Moliterni, A.G.G.; Polidori, G.; Spagna, R. SIR97: A new tool for crystal structure determination and refinement. *J. Appl. Crystallogr.* **1999**, *32*, 115–119. [CrossRef]

38. Kabuto, C.; Akine, S.; Nemoto, T.; Kwon, E.J.; Wakita, K. Yadokari-XG, Software for Crystal Structure Analyses, 2001; Release of Software (Yadokari-XG 2009) for Crystal Structure Analyses. *Cryst. Soc. Jpn.* **2010**, *51*, 218–224. [CrossRef]

 *magnetochemistry*

Article

Theoretical Study on Magnetic Interaction in Pyrazole-Bridged Dinuclear Metal Complex: Possibility of Intramolecular Ferromagnetic Interaction by Orbital Counter-Complementarity

Takuya Fujii [1], **Yasutaka Kitagawa** [1,2,]*, **Kazuki Ikenaga** [1], **Hayato Tada** [1], **Iori Era** [1] and **Masayoshi Nakano** [1,2,3,4,]*

[1] Department of Materials Engineering Science, Graduate School of Engineering Science, Osaka University, Toyonaka, Osaka 560-8531, Japan; tfujii@cheng.es.osaka-u.ac.jp (T.F.); kazuki.ikenaga@cheng.es.osaka-u.ac.jp (K.I.); hayato.tada@cheng.es.osaka-u.ac.jp (H.T.); iori.era@cheng.es.osaka-u.ac.jp (I.E.)
[2] Center for Spintronics Research Network (CSRN), Graduate School of Engineering Science, Osaka University, Toyonaka, Osaka 560-8531, Japan
[3] Quantum Information and Quantum Biology Division, Institute for Open and Transdisciplinary Research Initiatives, Osaka University, Toyonaka, Osaka 560-8531, Japan
[4] Institute for Molecular Science, 38 Nishigo-Naka, Myodaiji, Okazaki 444-8585, Japan
* Correspondence: kitagawa@cheng.es.osaka-u.ac.jp (Y.K.); mnaka@cheng.es.osaka-u.ac.jp (M.N.)

Received: 31 December 2019; Accepted: 24 February 2020; Published: 26 February 2020

Abstract: A possibility of the intramolecular ferromagnetic (FM) interaction in pyrazole-bridged dinuclear Mn(II), Fe(II), Co(II), and Ni(II) complexes is examined by density functional theory (DFT) calculations. When azide is used for additional bridging ligand, the complexes indicate the strong antiferromagnetic (AFM) interaction, while the AFM interaction becomes very weak when acetate ligand is used. In the acetate-bridged complexes, an energy split of the frontier orbitals suggests the orbital counter-complementarity effect between the d_{xy} orbital pair, which contributes to the FM interaction; however, a significant overlap of other d-orbital pairs also suggests an existence of the AFM interaction. From those results, the orbital counter-complementarity effect is considered to be canceled out by the overlap of other d-orbital pairs.

Keywords: pyrazole-bridged dinuclear metal complex; effective exchange integral (J); density functional theory (DFT); broken-symmetry (BS) method; orbital complementarity

1. Introduction

Since the discovery of single-molecule magnets (SMMs) [1–3], a huge number of studies have been devoted to finding the compounds with a higher blocking temperature (T_B). Because a blocking barrier height (U_{eff}) is in proportion to $|D|S^2$, where D and S are magnetic anisotropy and spin size, respectively, a larger (negative) D or larger S is required to realize the higher T_B. In recent years, for example, lanthanide ions such as Tb(III) or Dy(III) have been introduced to enlarge $|D|$ values by increasing the magnetic anisotropy. An alternative way to realize the high-T_B compound is an introduction of a plural number of metal ions to increase the spin size S. Because it is usually difficult to align the metal ions in the ferromagnetic manner, sometimes, the ferrimagnetic interaction using different valencies such as Mn(IV)/Mn(III) has been utilized to increase the size of S [3]. If we have rational and effective guidelines to align spins on each metal ion ferromagnetically, however, the size of S is easily increased. The ferromagnetic interaction is often explained by the orbital degeneracy due to the orbital orthogonality. The orbital orthogonality has succeeded in the explanation of many ferromagnetic materials [4]. For example, in the Prussian blue analogues (PBA), which is one of the ferromagnetic materials, the orbital degeneracy originating in the orthogonality between t_{2g} and e_g orbitals causes the ferromagnetic interaction [5–8]. The orbital orthogonality found in such PBA analogues, however, requires the linearly aligned metal–ligand–metal structures, as well as the specific combination of metal ions such as Cr(III)/Cr(II), so that it is considered to have less degrees of freedom in molecular structures. On the other hand, it has been known that some heterogeneous bridging ligands between two Cu(II) ions make different contributions to the magnetic exchange interactions [9,10]. A pyrazole, which is a heterocyclic five-membered ring compound, is often used as a bridging ligand between two metal ions [9–12]. Kida, Okawa, and co-workers reported that pyrazole-bridged dinuclear-Cu(II) complexes with azide and acetate as counter bridging-ligands shown in Figure 1 exhibit anti-ferromagnetic (AFM) and ferromagnetic (FM) behavior, respectively [9,10]. They explained this phenomenon by an orbital complementarity/counter-complementarity that originates in the orbital interaction between Cu(II) and bridging-ligands. On the other hand, the theoretical studies with ab initio and density functional theory (DFT) calculations have been one of the powerful tools for the investigation of the effective exchange integrals (J), especially for the polynuclear metal complexes [13–19]. Up to now, there have been only several papers that examine the exchange interactions of those complexes by theoretical calculations [14–18]; furthermore, those are devoted to elucidating the mechanism of the ferromagnetic interaction between two metal ions. The concept of the orbital counter-complementarity, however, predicts the ferromagnetic materials from differences in the phase of molecular orbitals between ligands and metals; therefore, it can be applied to the theoretical design of ferromagnetic molecules. As described below, it is considered that the magnetic interaction is not necessarily ferromagnetic, except for the Cu(II) complex, because of competition between the ferromagnetic interaction by orbital counter-complementarity and antiferromagnetic interaction by direct d–d overlap. In this study, as the first step for the molecular design, we examine whether this concept can be applied to metal ions, except for Cu(II), by introducing them into the model structures. Here, we examine the high-spin species of divalent 3d metal ions; Mn(II), Fe(II), Co(II), and Ni(II).

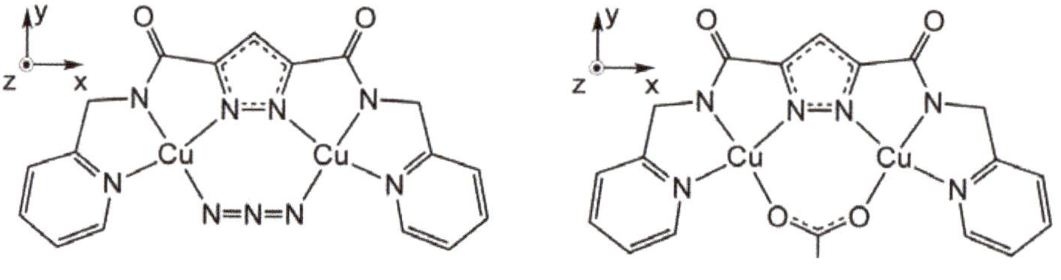

complex **1**

complex **2**

Figure 1. Dinuclear Cu(II) complexes bridged by azide (complex **1**) and acetate (complex **2**) ligands.

2. Theoretical Background

2.1. Orbital Complementarity and Counter-Complementarity

The concept of the orbital complementarity and counter-complementarity is briefly explained here using the dinuclear copper(II) complexes illustrated in Figure 1 [9,10]. Because each Cu(II) ion has a spin in the d_{xy} orbital, two Cu(II) ions in the complex form π-type symmetric bonding (φ_s) and anti-symmetric anti-bonding (φ_a) orbitals within the xy plane, as illustrated in Figure 2. The metal–metal bonding and anti-bonding orbitals also interact with the bridging-ligand orbitals. The pyrazole (φ_{pz}) ligand, which provides an anti-symmetric coordination orbital, interacts with the anti-symmetric (φ_a) orbital and forms in-phase ($\varphi_a + \varphi_{pz}$) and out-of-phase ($\varphi_a - \varphi_{pz}$) orbitals. In the case of complex **1**, the azide ligand orbital (φ_{az}) can interact with φ_a (Figure 2a), and the out-of-phase orbital ($\varphi^*_{a-(pz+az)}$) becomes unstable, while the bonding-orbital φ_s does not interact with ligand orbitals. As a result, two spins belong to the lower φ_s rather than $\varphi^*_{a-(pz+az)}$, and thus the AFM state is predicted to become the ground state. On the other hand, in the case of complex **2**, φ_a and φ_s interact with pyrazole (φ_{pz}) and acetate (φ_{ac}), respectively. Consequently, the two formed out-of-phase orbitals, that is, φ^*_{a-pz} and φ^*_{s-ac}, tend to be quasi-degenerate. Therefore, the FM state is predicted to become the ground state of complex **2**. Those explanations, however, assume an interaction among ligand orbitals and φ_a (and φ_s) orbitals that consist of d_{xy} of Cu(II). As mentioned above, we examine the high-spin species of Mn(II), Fe(II), Co(II), and Ni(II), because the low-spin species do not have spins in the d_{xy} orbital. Because metal ions examined in this paper have more spins (e.g., s = 5/2, 2, 3/2, and 1 for Mn(II), Fe(II), Co(II), and Ni(II), respectively), the direct overlap between other d-orbital pairs can contribute to the AFM interaction. In this sense, the relative stability between the FM and AFM states is predicted to be determined by a balance between the FM interaction via the orbital counter-complementarity caused by d_{xy} and the AFM interaction via direct overlap of other orbitals.

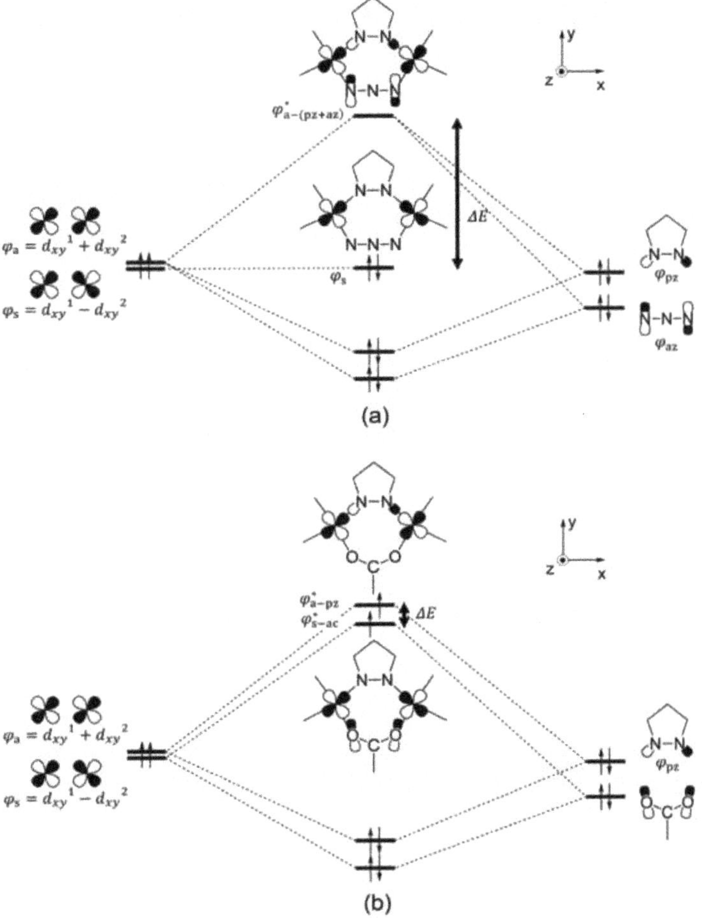

Figure 2. Schematic representation of orbital interactions for (**a**) complementarity and (**b**) counter-complementarity found in complexes **1** and **2**, respectively.

2.2. Computational Models

On the basis of the reported structures of two complexes **1** and **2** illustrated in Figure 1, we constructed model structures by substituting Mn(II), Fe(II), Co(II), and Ni(II) ions. In order to conduct a systematic analysis, we assume an octahedral coordination for all models and the axial positions of metal ions are capped by ammonia molecules that provide a moderate ligand field, as illustrated in Figure 3a. For comparison, Cr(II) complexes that do not have electrons in d_{xy} orbitals are also examined (Figure 3b).

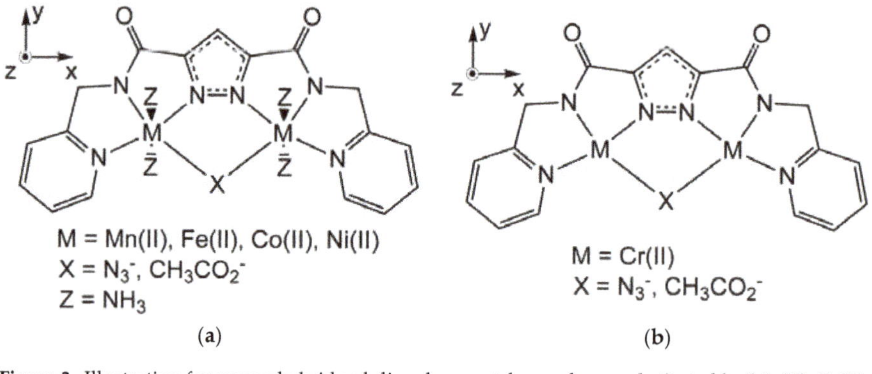

Figure 3. Illustration for pyrazole-bridged dinuclear metal complexes substituted by Mn(II), Fe(II), Co(II), Ni(II) ions (**a**), and Cr(II) ions (**b**). Note that a square planar structure is assumed for the Cr(II) complexes because the octahedral structure is not obtained by the geometry optimization.

2.3. Computational Details

For all calculations, we employed the broken-symmetry (BS) method [20] using the B3LYP functional [21] with the 6-31G* [22] basis set to obtain the spin-polarized electronic structures. The model structures were, at first, fully optimized and the optimized geometries were confirmed not to have any imaginary frequencies. In order to investigate the difference in the structures between the AFM and FM states, we performed the geometry optimization for both states. At each optimized geometry, the intramolecular magnetic interaction between two metal ions was discussed based on the effective exchange integral (J) values calculated by the Yamaguchi equation [23,24],

$$J = \frac{E^{\mathrm{AFM}} - E^{\mathrm{FM}}}{\langle S^2 \rangle^{\mathrm{FM}} - \langle S^2 \rangle^{\mathrm{AFM}}} \tag{1}$$

where E^{X} and $\langle S^2 \rangle^{\mathrm{X}}$ denote total energies and $\langle S^2 \rangle$ values of spin state X (X = FM and AFM). All calculations were performed in the gas phase using Gaussian09 [25].

3. Results and Discussion

3.1. Optimized Structures and Calculated J Values

The optimized Cartesian coordinates of model structures for both AFM and FM states are summarized in Table S1 in Supplementary Materials. The calculated total energies and $\langle S^2 \rangle$ values are summarized in Tables S2 and S3 in Supplementary Materials. We also confirm that the Fe(II) and Co(II) complexes energetically prefer the high-spin metal species rather than the low-spin species, as summarized in Tables S4 and S5 in Supplementary Materials. In addition, the stability of those model complexes is examined by roughly estimating the stabilization energies, as summarized in Table S6 in Supplementary Materials. The results indicate the possibility of existence of all model structures.

Calculated J values at the optimized geometry are summarized in Table 1. All models show negative J values even at the optimized structures for the FM states, indicating that the AFM exchange interaction between the metal ions is dominant even in the acetate-bridged complexes, in contrast to the Cu(II) complexes [17]. The azide-bridged complexes, however, exhibit significantly stronger AFM interaction, while the calculated J values of those acetate-bridged complexes are close to zero, indicating the small energy gap between the AFM and FM states. The result suggests that the orbital complementarity/counter-complementarity affects the magnetic interaction, like in Cu(II) complexes.

Table 1. Calculated J values at the optimized structures, where J_{AFM} and J_{FM} represent calculated J values with the antiferromagnetic (AFM) and ferromagnetic (FM) structures, respectively.

M	X	J_{AFM}/cm^{-1}	J_{FM}/cm^{-1}	Exptl [b]/cm^{-1}
Cr(II)	N_3^-	−8.9	−8.3	
	$CH_3CO_2^-$	−3.2	−3.1	
Mn(II)	N_3^-	−9.1	−8.1	
	$CH_3CO_2^-$	−0.3	−0.2	
Fe(II)	N_3^-	−13.1	−11.3	
	$CH_3CO_2^-$	−1.1	−1.0	
Co(II)	N_3^-	−26.1	−23.0	
	$CH_3CO_2^-$	−2.0	−0.1	
Ni(II)	N_3^-	−74.2	−33.1	
	$CH_3CO_2^-$	−3.2	−3.1	
Cu(II) [a]	N_3^-	−436	−364	−371
	$CH_3CO_2^-$	13.5	23.0	>8.9

[a] Calculated values in the work of [17]. [b] Experimental values in the work of [9].

3.2. Orbital Energy Difference

Next, we examined the energy gap between the symmetric and antisymmetric orbitals (ΔE), which is the origin of the concept of orbital complementarity, as illustrated in Figure 2. Here, ΔE is defined as follows,

$$\Delta E = \begin{cases} \left| E\left(\varphi^*_{a-(pz+az)}\right) - E(\varphi_s) \right| & X = N_3^- \\ \left| E\left(\varphi^*_{a-pz}\right) - E(\varphi^*_{s-ac}) \right| & X = CH_3CO_2^- \end{cases} \tag{2}$$

If ΔE value is negligible, those two orbitals shown in Figure 2 are quasi-degenerate and the FM state becomes stable owing to Hund's rule. As seen from Table 2, ΔE values of the azide-bridged complexes are larger than those of the acetate-bridged complexes. For example, the ΔE value of the acetate-bridged Mn(II) complex is 0.06 eV. ΔE values of the acetate-bridged complexes are shown to be small as compared with that of the Cu(II) complex [17]; therefore, the FM state is expected to become stable. However, their J values are found to be small, but still negative. This result indicates another mechanism to stabilize the AFM state, suggesting the overlap between metal ions via other d-orbitals.

3.3. Natural Orbital Analyses

In order to explain the AFM interaction of the acetate-bridged complexes, spin-unrestricted natural orbital (UNO) analysis is performed for the AFM state of each complex. The UNOs are obtained by diagonalizing the first order density matrix, and the occupation number of orbital i (n_i) is related to the overlap between α and β orbitals (T_i) as

$$T_i = n_i - 1. \tag{3}$$

If n_i is close to 1.0, then the two electrons in the corresponding α and β orbitals are spatially polarized, indicating that spins on the metal ions are almost localized there, that is, the fully spin-polarized state. On the other hand, n_i should be equal to 2.0 if those are fully overlapped, and spins disappear. The calculated occupation numbers of magnetic orbitals are summarized in

Table 3 and corresponding natural orbitals are depicted in Figures S1–S8. Although the d_{xy} orbitals of which the occupation numbers are close to 1.0 are spin-polarized, they still have significant overlap. In addition, other metal d-orbital pairs also have larger overlaps that contribute to the AFM interaction with the direct d-d overlap. For example, the overlaps between d_{zx} orbitals that are in a range of 0.013–0.025 is found to be comparable to those of d_{xy} in the acetate-bridged complexes. As a result, those orbital overlaps between those d-orbital pairs are considered to contribute to the stabilization of the AFM state. In those complexes, metal ions and bridging ligands are in the same plane; therefore, the metal–metal distance is a clue to consider the magnetic interaction. As summarized in Table S7 in Supplementary Information, the metal–metal distances depend on the metal species; however, they do not show a significant relationship between the overlap and the distances. It suggests that such a small overlap does not change the molecular structure, but only the magnetic interaction.

Table 2. Energy gap between the symmetric and antisymmetric combinations of the magnetic orbitals (ΔE) in the FM state.

M	X	ΔE / eV
Mn(II)	N_3^-	0.42
	$CH_3CO_2^-$	0.06
Fe(II)	N_3^-	0.40
	$CH_3CO_2^-$	0.07
Co(II)	N_3^-	0.29
	$CH_3CO_2^-$	0.12
Ni(II)	N_3^-	0.20
	$CH_3CO_2^-$	0.18
Cu(II) [a]	N_3^-	1.33
	$CH_3CO_2^-$	0.60

[a] In the work of [17]. Note that the values are calculated by UBHandHLYP/6-31G* level of theory, and thus overestimated.

Table 3. Occupation numbers (n) of unrestricted natural orbitals (UNOs) for magnetic orbitals of model complexes, where dominant d-orbitals in each UNO are expressed in square parentheses.

M	X	HONO	HONO-1	HONO-2	HONO-3	HONO-4
Mn(II)	N_3^-	1.002 $\left[d_{x^2-y^2}\right]$	1.011 $\left[d_{yz}\right]$	1.012 $[d_{z^2}]$	1.025 $[d_{zx}]$	1.093 $\left[d_{xy}\right]$
	$CH_3CO_2^-$	1.006 $[d_{z^2}]$	1.007 $\left[d_{x^2-y^2}\right]$	1.013 $\left[d_{yz}\right]$	1.017 $\left[d_{xy}\right]$	1.018 $[d_{zx}]$
Fe(II)	N_3^-	1.009 $\left[d_{yz}\right]$	1.012 $[d_{z^2}]$	1.021 $[d_{zx}]$	1.087 $\left[d_{xy}\right]$	
	$CH_3CO_2^-$	1.004 $\left[d_{x^2-y^2}\right]$	1.009 $\left[d_{yz}\right]$	1.016 $[d_{zx}]$	1.019 $\left[d_{xy}\right]$	
Co(II)	N_3^-	1.011 $[d_{z^2}]$	1.023 $[d_{zx}]$	1.094 $\left[d_{xy}\right]$		
	$CH_3CO_2^-$	1.007 $[d_{z^2}]$	1.013 $[d_{zx}]$	1.022 $\left[d_{xy}\right]$		
Ni(II)	N_3^-	1.019 $[d_{z^2}]$	1.113 $\left[d_{xy}\right]$			
	$CH_3CO_2^-$	1.012 $[d_{z^2}]$	1.029 $\left[d_{xy}\right]$			
Cu(II)[a]	N_3^-	1.188 $\left[d_{xy}\right]$				
	$CH_3CO_2^-$	1.002 $\left[d_{xy}\right]$				

[a] Calculated values in the work of [17].

In order to confirm the effects of the orbital complementarity, counter-complementarity, and direct overlap, we examined the Cr(II) complexes. As the high-spin Mn(II) ion has d^5 configuration, each d-orbital is filled by one electron in the complexes, while a d_{xy} orbital is not occupied in the case of the Cr(II) complex (d^4 configuration) owing to the strong ligand field by surrounding ligands. A difference in the calculated J values between azide- and acetate-bridged complexes for the Cr(II) complexes is found to be less than that for the Mn(II) complexes. This contrast suggests that the orbital counter-complementarity effect contributes to a decrease of the energy gap between the AFM and FM states. In addition, calculated J values of the azide-bridged Mn(II) and Cr(II) complexes are almost the same; nevertheless, the acetate-bridged Cr(II) complex shows a negative value about 3 cm^{-1} smaller than the Mn(II) complex. An electron configuration of the Cr(II) complexes can be obtained by eliminating electrons from d_{xy} orbitals of the Mn(II) complexes; therefore, roughly speaking, the difference is predicted to originate in the stabilization of the FM state caused by the degeneracy in φ^*_{a-pz} and φ^*_{s-ac}, that is, orbital counter-complementarity. From those results, it is predicted that the FM interaction via the counter-complementarity is canceled by the AFM interaction via overlap of other d-orbitals in those complexes.

Finally, we consider a possibility of an end-on type coordination of the azide complex. It has been reported that the azide ligand can bridge two metal ions with the end-on type manner as well as the side-on type, which is considered above [26,27]. We performed geometry optimization of the dinuclear Cu(II) complex for the end-on type structure, which is summarized in Table S8 in Supplementary Information. The total energies of the side-on and end-on type structures that are summarized in Table S9 in Supplementary Information indicate the side-on type is significantly stable in comparison with the end-on type. Therefore, we did not mention the end-on type structures of other metal complexes in this manuscript, although the end-on type induces the ferromagnetic interaction.

4. Conclusions

In this paper, we examine the possibility of the intramolecular FM interaction in pyrazole-bridged dinuclear Mn(II), Fe(II), Co(II), and Ni(II) complexes. We confirmed that the azide-bridged complexes indicate the strong AFM interaction, while the acetate-bridged complexes exhibit a very weak AFM interaction. These results suggest that the counter-complementarity effect, which contributes to the FM interaction, competes with the AFM interaction caused by overlap of other d-orbital pairs. Conversely, there is the possibility that these 3d metal complexes exhibit an FM interaction if one can decrease the overlap. For that purpose, a rational designing guideline based on the quantum chemical calculation is considered to be effective for modifying the species of substituents and their substitution positions [28].

Supplementary Materials: The following are available online at http://www.mdpi.com/2312-7481/6/1/10/s1, Table S1: Optimized cartesian coordinates (Å) of acetate bridged dinuclear Ni(II) complex; Table S2: Calculated total energies and $\langle S^2 \rangle$ values at AFM structures; Table S3: Calculated total energies and $\langle S^2 \rangle$ values at FM structures; Table S4: Calculated total energies and $\langle S^2 \rangle$ values of low-spin Fe(II) ($s = 0$) complexes; Table S5: Calculated total energies and $\langle S^2 \rangle$ values of low-spin Co(II) ($s = 1/2$) complexes; Table S6: Stabilization energy (kJ/mol) for each complex; Table S7: Metal-metal distances (Å) in each complex; Table S8: Optimized cartesian coordinates (Å) of end-on azide bridged dinuclear Cu(II) complex; Table S9: Calculated total energies and $\langle S^2 \rangle$ values at AFM structures for dinuclear Cu(II) complex with side-on and end-on type azide bridges; Figure S1: Calculated natural orbitals of magnetic orbitals for acetate-bridged Mn(II) complex; Figure S2: Calculated natural orbitals of magnetic orbitals for acetate-bridged Fe(II) complex; Figure S3: Calculated natural orbitals of magnetic orbitals for acetate-bridged Co(II) complex; Figure S4: Calculated natural orbitals of magnetic orbitals for acetate-bridged Ni(II) complex; Figure S5: Calculated natural orbitals of magnetic orbitals for azide-bridged Mn(II) complex; Figure S6: Calculated spin-unrestricted natural orbitals of magnetic orbitals for azide-bridged Fe(II) complex; Figure S7: Calculated spin-unrestricted natural orbitals of magnetic orbitals for azide-bridged Co(II) complex; Figure S8: Calculated spin-unrestricted natural orbitals of magnetic orbitals for azide-bridged Ni(II) complex.

Author Contributions: Y.K. and T.F. designed the study and wrote the paper. K.I., H.T., and I.E. contributed calculations. M.N. organized the study. All authors have read and agreed to the published version of the manuscript.

Funding: This work was supported by Grant-in-Aid for Scientific Research (KAKENHI) (Nos. 19K05401, 26410093, and JP18H01943) from Japan Society for the Promotion of Science (JSPS). This work was also performed under the Inter-University Cooperative Research Program of the Institute for Materials Research, Tohoku University (Proposal No. 19K0061).

Conflicts of Interest: The authors declare no conflict of interest.

References

1. Caneschi, A.; Gatteschi, D.; Sessoli, R.; Barra, A.L.; Brunel, L.C.; Guillot, M. Alternating current susceptibility, high field magnetization, and millimeter band EPR evidence for a ground S = 10 state in [Mn$_{12}$O$_{12}$(CH$_3$COO)$_{16}$(H$_2$O)$_4$]·2CH$_3$COOH·4H$_2$O. *J. Am. Chem. Soc.* **1992**, *113*, 5873–5874. [CrossRef]
2. Sessoli, R.; Gatteschi, D.; Caneschi, A.; Novak, M.A. Magnetic bistability in a metal-ion cluster. *Nature* **1993**, *365*, 141–143. [CrossRef]
3. Christou, G.; Gatteschi, D.; Hendrickson, D.N.; Sessoli, R. Single-Molecule Magnets. In *MRS Bulletin*; Cambridge University Press: Cambridge, UK, 2000; Volume 25, pp. 66–71.
4. Oshio, H.; Nihei, M.; Yoshida, A.; Nojiri, H.; Nakano, M.; Yamaguchi, A.; Karaki, Y.; Ishimoto, H. A Dinuclear MnIII–CuIISingle-Molecule Magnet. *Chem. Euro. J.* **2005**, *11*, 843–848. [CrossRef] [PubMed]
5. Gadet, V.; Mallah, T.; Castro, I.; Verdaguer, M.; Veillet, P. High-T$_C$ molecular-based magnets: A ferromagnetic bimetallic chromium(III)-nickel(II) cyanide with T$_C$ = 90 K. *J. Am. Chem. Soc.* **1992**, *114*, 9213–9214. [CrossRef]
6. Mallah, T.; Thiébaut, S.; Verdaguer, M.; Veillet, P. High-T$_c$ Molecular-Based Magnets: Ferrimagnetic Mixed-Valence Chromium(III)-Chromium(II) Cyanides with T$_c$ at 240 and 190 Kelvin. *Science* **1993**, *262*, 1554–1557. [CrossRef] [PubMed]
7. Tokoro, H.; Ohkoshi, S. Novel magnetic functionalities of Prussian blue analogs. *Dalton Trans.* **2011**, *40*, 6825–6833. [CrossRef] [PubMed]
8. Baker, M.L.; Kitagawa, Y.; Nakamura, T.; Tazoe, K.; Narumi, Y.; Kotani, Y.; Iijima, F.; Newton, G.N.; Okumura, M.; Oshio, H.; et al. X-ray Magnetic Circular Dichroism Investigation of the Electron Transfer Phenomena Responsible for Magnetic Switching in a Cyanide-Bridged [CoFe] Chain. *Inorg. Chem.* **2013**, *52*, 13956–13962. [CrossRef] [PubMed]
9. Kamiusuki, T.; Okawa, H.; Kitaura, E.; Koikawa, M.; Matsumoto, N.; Kida, S.; Oshio, H. Binuclear copper(II) complexes of new dinucleating ligands with a pyrazolate group as an endogenous bridge. Effects of exogenous azide and acetate bridges on magnetic properties. *J. Chem. Soc. Dalton Trans.* **1989**, 2077–2081. [CrossRef]
10. Nishida, Y.; Kida, S. An Important Factor Determining the Significant Difference in Antiferromagnetic Interactions between Two Homologous (μ-Alkoxo)(μ-pyrazolato-*N,N'*)dicopper(II) Complexes. *Inorg. Chem.* **1988**, *27*, 447–452. [CrossRef]
11. Behle, L.; Neuburger, M.; Zehnder, M.; Kaden, T.A. Metal Complexes with Macrocyclic Ligands. Part XXXIX. Mono- and binuclear copper(II) complexes of a bridging bis[1,4,7-triazacyclononane]. *Helv. Chim. Acta* **1995**, *78*, 693–702. [CrossRef]
12. Meyer, F.; Beyreuther, S.; Heinze, K.; Zsolnai, L. Dinuclear Cobalt(II) Complexes of Pyrazole Ligands with Chelating Side Arms. *Eur. J. Inorg. Chem.* **1997**, *130*, 605–613. [CrossRef]
13. Pons, J.; López, X.; Benet, E.; Casabó, J.; Teixidor, F.; Sánchez, F.J. Dinuclear μ-pyrazole nickel(II), cobalt(II), cadmium(II) and zinc(II) complexes with dinucleating pyrazole-derived ligands. *Polyhedron* **1990**, *23*, 2839–2845. [CrossRef]
14. Kitagawa, Y.; Saito, T.; Yamaguchi, K. Approximate Spin Projection for Broken-Symmetry Method and Its Application. In *Symmetry (Group Theory) and Mathematical Treatment in Chemistry*; Akitsu, T., Ed.; IntechOpen: London, UK, 2018. [CrossRef]
15. Tugrul Zeyrek, C. Importance of Orbital Complementarity in Spin Coupling through Two Different Bridging Groups in Dicopper(II) Complexes of Endogenous Alkoxo Bridging Ligand with Exogenous Carboxylate: Ab-initio and Semi-Empirical Calculations. *Z. Naturforsch* **2007**, *62*, 409–416. [CrossRef]
16. Wang, L.-L.; Sun, Y.-M.; Qui, Z.-N.; Liu, C.-B. Magnetic Interactions in Two Heterobridged Dinuclear Copper(II) Complexes: Orbital Complementarity or Countercomplementarity? *J. Phys. Chem. A* **2008**, *112*, 8418–8422. [CrossRef] [PubMed]
17. Miyagi, K.; Kitagawa, Y.; Asaoka, M.; Teramoto, R.; Natori, Y.; Saito, T.; Nakano, M. Theoretical study of magnetic interaction in pyrazole-bridged dinuclear Cu(II) complex. *Polyhedron* **2017**, *136*, 132–135. [CrossRef]
18. Kachi-Terajima, C.; Ishii, R.; Tojo, Y.; Fukuda, M.; Kitagawa, Y.; Asaoka, M.; Miyasaka, H. Ferromagnetic Exchange Coupling in a Family of Mn^{III} Salen-Type Schiff-Base Out-of-Plane Dimers. *J. Phys. Chem. C* **2017**, *121*, 12454–12468. [CrossRef]
19. Wang, L.-L.; Sun, Y.-M.; Gao, J.; Lin, X.-J.; Liu, C.-B. Insights into the control of magnetic coupling in the Mn$_4$^{III} complex: From ferromagnetic to antiferromagnetic. *Dalton Trans.* **2010**, *39*, 10249–10255. [CrossRef]

20. Fukutome, H. The Unrestricted Hartree-Fock Theory of Chemical Reactions. I: The Electronic Instabilities in the Chemical Reactions and the Solutions of the Unrestricted SCF LCAO MO Equation for the Homopolar Two-Center Two-Electron System. *Prog. Theoret. Phys.* **1972**, *47*, 1156–1180. [CrossRef]
21. Becke, A. Density-functional thermochemistry. III. The role of exact exchange. *J. Chem. Phys.* **1993**, *98*, 5648–5652. [CrossRef]
22. Ditchfield, R.; Hehre, W.J.; Pople, J.A. Self-Consistent Molecular Orbital Methods. IX. Extended Gaussian-type basis for molecular-orbital studies of organic molecules. *J. Chem. Phys.* **1971**, *54*, 724–728. [CrossRef]
23. Yamaguchi, K.; Fukui, H.; Fueno, T. Molecular orbital (MO) theory for magnetically interacting organic compounds. Ab-initio MO calculations of the effective exchange integrals for cyclophane-type carbene dimers. *Chem. Lett.* **1986**, *15*, 625–628. [CrossRef]
24. Soda, T.; Kitagawa, Y.; Onishi, T.; Takano, Y.; Shigeta, Y.; Nagao, H.; Yoshioka, Y.; Yamaguchi, K. Ab initio computations of effective exchange integrals for H–H, H–He–H and Mn_2O_2 complex: Comparison of broken-symmetry approaches. *Chem. Phys. Lett.* **2000**, *319*, 223–230. [CrossRef]
25. Frisch, M.J.; Trucks, G.W.; Schlegel, H.B.; Scuseria, G.E.; Robb, M.A.; Cheeseman, J.R.; Scalmani, G.; Barone, V.; Mennucci, B.; Petersson, G.A.; et al. *Gaussian 09, Revision C.01*; Gaussian, Inc.: Wallingford, CT, USA, 2009.
26. Yoon, J.H.; Ryu, D.W.; Kim, H.C.; Yoon, S.W.; Suh, B.J.; Hong, C.S. An End-On Azide-Bridged Antiferromagnetic Single-Chain Magnet Involving Spin Canting and Field-Induced Two-Step Magnetic Transitions. *Chem. Enro. J.* **2009**, *15*, 3661–3665. [CrossRef] [PubMed]
27. Ge, C.-H.; Cui, A.-L.; Ni, Z.-H.; Jiang, Y.-B.; Zhang, L.-F.; Ribas, J.; Kou, H.-Z. $\mu_{1,1}$-Azide-Bridged Ferromagnetic Mn^{III} Dimer with Slow Relaxation of Magnetization. *Inorg. Chem.* **2006**, *45*, 4883–4885. [CrossRef] [PubMed]
28. Natori, Y.; Kitagawa, Y.; Aoki, S.; Teramoto, R.; Tada, H.; Era, I.; Nakano, M. Quantum chemical design guidelines for absorption and emission color tuning of *fac*-Ir(ppy)$_3$ complexes. *Molecules* **2018**, *23*, 577. [CrossRef] [PubMed]

MDPI
St. Alban-Anlage 66
4052 Basel
Switzerland
Tel. +41 61 683 77 34
Fax +41 61 302 89 18
www.mdpi.com

Magnetochemistry Editorial Office
E-mail: magnetochemistry@mdpi.com
www.mdpi.com/journal/magnetochemistry